Practical Elements of Safety

Springer
London
Berlin
Heidelberg
New York
Hong Kong
Milan
Paris
Tokyo

Related titles:

Towards System Safety
Proceedings of the Seventh Safety-critical Systems Symposium, Huntingdon, UK 1999
Redmill and Anderson (Eds)
1-85233-064-3

Lessons in System Safety
Proceedings of the Eighth Safety-critical Systems Symposium, Southampton, UK 2000
Redmill and Anderson (Eds)
1-85233-249-2

Aspects of Safety Management
Proceedings of the Ninth Safety-critical Systems Symposium, Bristol, UK 2001
Redmill and Anderson (Eds)
1-85233-411-8

Components of System Safety
Proceedings of the Tenth Safety-critical Systems Symposium, Southampton, UK 2002
Redmill and Anderson (Eds)
1-85233-561-0

Current Issues in Safety-critical Systems
Proceedings of the Eleventh Safety-critical Systems Symposium, Bristol, UK 2003
Redmill and Anderson (Eds)
1-85233-696-X

Felix Redmill and Tom Anderson (Eds)

Practical Elements of Safety

Proceedings of the Twelfth Safety-critical Systems Symposium, Birmingham, UK 17-19 February 2004

Safety-Critical
Systems Club

BAE SYSTEMS

Springer

Felix Redmill
Redmill Consultancy, 22 Onslow Gardens, London, N10 3JU

Tom Anderson
Centre for Software Reliability, University of Newcastle,
Newcastle upon Tyne, NE1 7RU

British Library Cataloguing in Publication Data
A catalogue record for this book is available from the British Library

Library of Congress Cataloging-in-Publication Data
Safety-Critical Systems Symposium (12th : 2004 : Birmingham, England)
Practical elements of safety : proceedings of the twelfth Safety-Critical Systems Symposium, Birmingham, UK, 17-19 February 2004 / Felix Redmill amd Tom Anderson (eds.).
p. cm.
"Safety-Critical Systems Club"
Includes bibliographical references.
ISBN 1-85233-800-8 (alk. Paper)
1. Industrial safety--Management--Congresses. 2. Automatic control--Reliability--Congresses. 3. Computer software--Reliability--Congresses. I. Redmill, Felix. II. Anderson, Tom, 1947- III. Title.
T55.A1S225 2004
363.11--dc21

2003067387

ISBN 1-85233-800-8 Springer-Verlag London Berlin Heidelberg
Springer-Verlag is a part of Springer Science+Business Media
springeronline.com

Printed in Great Britain

Typesetting: Camera ready by contributors
Printed and bound by the Athenæum Press Ltd., Gateshead, Tyne & Wear
34/3830-543210 Printed on acid-free paper SPIN 10951460

PREFACE

The practice of modern safety engineering and management is based on risk - on the identification and analysis of risks, the assessment of their tolerability and what should be done about them, the implementation of necessary risk reduction and mitigation, and the assessment of the entire process to ensure that each step was appropriate in the circumstances. But, straightforward as these activities may seem, they carry complexity, difficulty, and subjective judgement. The papers in this book - presented at the twelfth Safety-critical Systems Symposium - address issues integral to the practice of carrying them out.

Four papers address the handling of risk: two expose methods used in the UK railway industry to the wider community, one integrates the human contribution to risk with safety culture, and one explores the delicate interface between risk practitioners and the news media. Out of the understanding of the risks emerge safety requirements, and two papers illuminate the tricky and dimly lit area of safety integrity levels. For many years the proponents of formal methods have advocated their use in system specification and development, while sceptics have pointed to their obscurity and cost. Three papers claim that formal methods are now ready for industrial use, showing that with judicious application their benefits can be obtained easily and cost-effectively. Other papers expand on the derivation of evidence for the demonstration of safety, the development of a safety case, the assessment of safety, and the changing face of safety legislation, within the context of which all safety-critical-system activities are set.

The papers cover a broad spectrum and take a practical perspective. Of the fourteen, nine are from industry, one from a regulator, and one from academe. Three have both industrial and academic input. We thank the authors for their contributions and their co-operation in the preparation of this book - for sponsorship of which we also thank BAE Systems.

The Safety-critical Systems Symposium, organised by the Safety-Critical Systems Club, provides a forum for inter-disciplinary technology transfer. Because it does so in a 'club' atmosphere rather than a commercial environment, problems are more readily admitted to and lessons more openly communicated. We thank Joan Atkinson for the logistical organisation that makes this possible.

FR and TA
October 2003

THE SAFETY-CRITICAL SYSTEMS CLUB

sponsor and organiser
of the

Safety-critical Systems Symposium

What is the Club?

The Safety-Critical Systems Club exists to raise awareness of safety issues and to facilitate technology transfer in the field of safety-critical systems. It is an independent, non-profit organisation that co-operates with all bodies involved with safety-critical systems.

History

The Club was inaugurated in 1991 under the sponsorship of the Department of Trade and Industry (DTI) and the Engineering and Physical Sciences Research Council (EPSRC). Its secretariat is at the Centre for Software Reliability (CSR) in the University of Newcastle upon Tyne, and its Co-ordinator is Felix Redmill of Redmill Consultancy.

Since 1994 the Club has had to be self-sufficient, but it retains the active support of the DTI and EPSRC, as well as that of the Health and Safety Executive, the Institution of Electrical Engineers, and the British Computer Society. All of these bodies are represented on the Club's Steering Group.

What does the Club do?

The Club achieves its goals of technology transfer and awareness-raising by focusing on current and emerging practices in safety engineering, software engineering, and standards that relate to safety in processes and products. Its activities include:

- Running the annual Safety-critical Systems Symposium each February (the first was in 1993), with Proceedings published by Springer-Verlag;
- Organising a number of 1- or 2-day seminars each year;
- Providing tutorials on relevant subjects;
- Publishing a newsletter, *Safety Systems*, three times each year (since 1991), in January, May and September.

How does the Club help?

The Club brings together technical and managerial personnel within all sectors of the safety-critical community. It provides education and training in principles and techniques, and facilitates the dispersion of lessons within and between industry sectors. It promotes an inter-disciplinary approach to safety engineering and management and provides a forum for experienced practitioners to meet each other and for the exposure of newcomers to the safety-critical systems industry.

The Club facilitates communication among researchers, the transfer of technology from researchers to users, feedback from users, and the communication of experience between users. It provides a meeting point for industry and academia, a forum for the presentation of the results of relevant projects, and a means of learning and keeping up-to-date in the field.

The Club thus helps to achieve more effective research, a more rapid and effective transfer and use of technology, the identification of best practice, the definition of requirements for education and training, and the dissemination of information. And it does this within a 'club' atmosphere rather than a commercial environment.

Membership

Members pay a reduced fee (well below a commercial level) for events and receive the newsletter and other mailed information. Without sponsorship, the Club depends on members' subscriptions, which can be paid at the first meeting attended.

To join, please contact Mrs Joan Atkinson at: Centre for Software Reliability, University of Newcastle upon Tyne, NE1 7RU; Telephone: 0191 221 2222; Fax: 0191 222 7995; Email: csr@newcastle.ac.uk

CONTENTS LIST

THE HUMAN SIDE OF RISK

ASSESSMENT AND THE DERIVATION OF EVIDENCE

SAFETY ARGUMENT AND THE LAW

KEYNOTE

Dear Sir, Yours faithfully: an Everyday Story of Formality

Peter Amey

Praxis Critical Systems, 20, Manvers St., Bath BA1 1PX, UK
peter.amey@praxis-cs.co.uk

Abstract. The paper seeks a perspective on the *reality* of Formal Methods in industry today. What has worked; what has not; and what might the future bring? We show that where formality has been adopted it has largely been benefical. We show that formality takes many forms, not all of them obviously "Formal Methods".

1 Introduction

This could have been the shortest paper in the history of the SCSC. Formal Methods were briefly promoted by a small group of academic zealots in the 1980s. There were no useable tools, the methods didn't scale to real problems and industry—after careful evaluation of the evidence—quickly abandoned them. Formal methods are now dead (more dead even than Ada, if that's possible). End of story.

Not true, certainly, but a pastiche that is not a million miles from the perceptions, and prejudices, of much of our industry. These misconceptions and myths were well expressed by Anthony Hall over 10 years ago [Hall 1990]; many of his observations are as fresh and relevant as if they were written yesterday. The truth about Formal Methods, as always, lies somewhere between the extremes expressed by pundit and critic. Formal methods have not proved to be a panacea (although this claim was usually a straw man erected by critics rather than something seriously advanced by enthusiasts). Neither have Formal Methods died. In fact they are widely used; sometimes in a rather niche way; sometimes so routinely that they no longer attract attention, and sometimes in disguise.

I hope to show that formality, whether or not it comes with the label "Formal Methods" is alive and well and has a future. Indeed, if our industry is to rise to the challenges of the 21st century, mathematical rigour, in some form, will have an essential role to play (as it has in *all* other engineering disciplines).

For the rest of this paper I propose to abandon the capitalization of Formal Methods. In any case, this promotion to proper noun status is, I think, harmful in that it somehow labels normal good engineering as "special", "unusual" or just plain "different". I will use the term *formal* simply to mean *underpinned by mathematical rigour* (whether or not the mathematics is visible to the user). A formal method (without the capitals) simply means a *method underpinned by mathematical rigour*.

2 An Historical Overview

The need for precise ways to interact with the fast but stupid machines we call computers was recognised early in their life. In a 1948 University of Manchester paper, Alan Turing noted "one could communicate with these machines in any language provided it was an exact language" and "the system should resemble normal mathematical procedure closely, but at the same time should be as unambiguous as possible" [Hodges 1992]. Despite this early wisdom, the initial priority was for productivity rather than precision. It was only at the beginning of the 1970s that it became clear that the growing power of hardware and the availability of high-level languages such as FORTRAN, meant that it was possible to construct systems that exceeded our capabilities for specification, verification and validation. In his 1972 Turing Award lecture, Dijkstra observed: "The vision is that, well before the 1970s have run to completion, we shall be able to design and implement the kind of systems that are now straining our programming ability at the expense of only a few percent in man-years of what they cost us now, and that besides that, these systems will be virtually free of bugs". I think, in defence of Dijkstra's reputation, we should make it clear this was a *vision* not a *prediction!* We should also be clear, as observed recently by Martyn Thomas [Thomas 2003], that 30 years after that lecture, we have not even come close to achieving Dijkstra's vision.

The Formal Methods (capitals intentional here) movement that really got under way in the 1980s was just one of the responses to the growing software crisis. Other responses included the development of Ada by the US DoD, to replace the polyglot chaos that they then endured (and are now rapidly returning to); the adoption of static analysis by UK MoD certification agencies; and the production of new, tighter standards such as Def Stan 00-55. My overwhelming recollection of each of these events is the hostility they generated in much of the software industrial base. I particularly recall the launch event and panel session at RSRE Malvern for Interim Def Stan 00-55 where industry representatives queued up to explain that it wouldn't work, couldn't be made to work and shouldn't even be tried while at the same time remaining silent on how else we might advance the software engineering cause.

These perversities may be explained by the "guru effect". Since we have imperfect processes for producing software, we often rely on the exceptional skill of a very few people to get things to work. These are people who, amongst other things, happen to be able to work effectively with our imperfect processes. These people soon get a reputation for their skill. When alternative approaches are mooted, to whom does the manager turn for advice and guidance? These same experts. Are these experts likely to recommend the adoption of techniques that may undermine their status as company guru? No. They may recommend change but only to something they are comfortable with or which they find cool or interesting. In this sense, the company-saving guru may also be its biggest obstacle to progress.

3 No Evidence?

Before describing the current situation and laying down some hostages to fortune for the future, I would like briefly to address the suggestion that industry decided not to adopt formal methods because a careful business analysis revealed little advantage in it. If only that were true. I would even be happy to stop promoting precision, rigour and formality if it were. The truth is that:

- our industry is highly prone to rejecting or adopting technologies without *any* kind of analysis of their merits (witness the rush to object oriented methods or to the UML); and
- such evidence as is available shows the adoption of formal methods to be very largely advantageous.

As a motivator there follows a brief selection of formal methods successes. Note that these are *business* successes as well as *technical* successes.

3.1 CDIS

The CCF (Central Control Function) Display Information System is rather an old project now, it was delivered by Praxis to the UK CAA (Civil Aviation Authority) in 1992. It rates a mention here because of relative novelty of the approach (at the time) and the excellence of the results achieved. An abstract VDM model was produced along with more concrete user interface definitions. Code was produced from progressive informal refinements of the abstract VDM model. The high-availability, dual LAN was specified using CCS (Calculus of Communicating Systems). See [Hall 1996] for a more detailed descrition of this project. End-to-end productivity was the same or better than on comparable projects using informal approaches but the delivered software had a defect rate of about 0.75 faults per kloc, approaching an order of magnitude better than comparable ATC (Air Traffic Control) systems. This extra quality was free. The CDIS system has been exceptionally trouble free in over 10 years service.

3.2 Lockheed C130J

The Lockheed C130J mission computer software is a good example of formality by stealth. The software specification was written in a tabular, functional form using the Software Productivity Consortium's CORE notation [SPC 1993]. The code itself was written in SPARK [Barnes 2003] and SPARK's annotations were used to bind the specification and code tightly together. The CORE specification looks remarkably unthreatening but actually has clearly-defined semantics allowing the automatic generation of test cases from it. SPARK is also rigorously defined and leaves no hiding place for vagueness and ambiguity; indeed, a key benefit from its use was the way it forced coders to challenge anything unclear in the specification they were being asked to implement.

It might be thought that Lockheed were prepared to accept the obvious pain that CORE and SPARK must have inflicted because of the extreme criticality of the software under development. Actually, this view is quite wrong: there was no pain. In fact the code quality improved (by one or two orders of magnitude over comparable flight critical software) and the cost of development fell (to a quarter of that expected)[Amey 2002]. The savings arose from the reduction in the very expensive testing process associated with achieving Level A assurance against DO-178B [RTCA 1992] and, in particular, the virtual elimination of retesting caused by the detection and correction of bugs. Again we find that formality applied to the development of systems both raises quality and reduces cost.

3.3 SHOLIS

The Ship Helicopter Operating Limits Information System provides existential proof that Interim Def Stan 00-55 was a practical standard and could be followed without exorbitant cost and with good results (even with tools and with computing power that fall well short of their current equivalents). SHOLIS is a safety-related system that determines whether a particular helicopter manoeuvre is safe for a given ship in a particular sea and wind state. SHOLIS was specified in Z, coded in SPARK and correctness proofs (technically *partial* proofs since termination was not proved) used to bind the code and specification together. A survey of the project was carried out in conjunction with the University of York [King 2000]. The project illustrates both the power and the limitations of the formal methods used. On the positive side the proof process was both tractable and cost-effective. In particular, proof effort on the specification, before any coding effort had been expended, revealed many subtle flaws that could have emerged later and been more expensive to correct. Overall, proof—in terms of errors eliminated per man hour expended—was more effective than traditional activities such as unit test. In fact the project provides good evidence for abandoning comprehensive unit testing on projects where strong notations such as SPARK and are used—a significant saving. No errors have been reported in the SHOLIS system in sea trials and initial deployment.

Less positively, the system did experience (a few) late-breaking problems. Some of these were found during acceptance testing and involved requirements that were outside the formal model of the system represented by the specification. For example, there was a requirement for the system to tolerate the removal and replacement of arbitrary circuit boards during operation, something that clearly cannot be specified in Z. The main lesson here is to understand what is within and what is without (outwith for Scottish readers) the formal system model and to take adequate verification and validation steps for the latter.

3.4 MULTOS CA

Praxis Critical Systems developed the Certification Authority (CA) for the MULTOS [MULTOS] smart card scheme on behalf of Mondex International. The approach taken is detailed in [Hall 2002a] and [Hall 2002b]. Unlike some of the other example projects the MULTOS CA is *security critical* rather than *safety critical* and was developed to meet the requirements of ITSEC E6, a security classification broadly equivalent to SIL4 in the safety world [ITSEC 1991]. The system was COTS based and incorporated C++ (for the user interface GUI); C (for interfaces to specialized encryption hardware); an SQL database; Ada 95; and SPARK (for the key security-critical functions).

The system specification was produced in Z and, with the full agreement of the customer, was made the definitive arbiter of desired system behaviour. This agreement was significant because Praxis offered a warranty for the system where any deviation from the specified behavour would be fixed at Praxis's expense; this would have been much more difficult to agree if there was no rigorous description of the system's expected behaviour and therefore no common ground on which to agree whether a warranty claim was justified.

A particularly striking feature of this project is the way that errors were usually detected very close, in time, to the point where they were introduced (see [Hall 2002b]). There were very few cases where an error introduced early in the lifecycle (for example, in the specification) remained undetected until late in the lifecycle (for example, integration testing). This is a key property that distinguishes systems developed formally (which can be *reasoned about*) from those developed in a more ad hoc manner (and which can only be *tested*). A consequence of this early error elimination—the very essence of the term "correctness by construction"—is high productivity and low residual error rates. The MULTOS CA delivered an end-to-end productivity—including requirements, testing and management—of 28 lines of code per man day which is high for a SIL4 equivalent project. This high productivity was coupled with a residual error rate, corrected under warranty, of 0.04 defects per KSLOC (about 250% better than the space shuttle software! [Keller 1993]).

Of particular note is the fact that the customer, who now maintains the system, has adopted and maintained the formal specification because of its obvious value; perhaps as a financial organization they are showing greater wisdom and less prejudice than the software world?

3.5 TCAS

Most, if not all, large commercial aircraft are now equipped with the Traffic Collision Avoidance System (TCAS). Interestingly, the official specification for the TCAS II system, sponsored by the US Federal Aviation Administration, is a formal one, written in RSML (Requirements State Machine Language). Prior to its adoption, the formal specification was produced by the Safety-Critical Systems Research Group at the University of California, Irvine. A parallel effort by an industry group to produce an English specification was abandoned because of the difficulty of coping with the complexity of the TCAS function using an informal notation.

The adoption of the RSML specification has had some important benefits. In particular, it has been possible to check it for mathematical completeness and consistency [Heimdahl 1996]. The widespread use of the specification has also spawned a number of supporting tools including code generators, test-case generators and simulators; none of these would have been possible using an informal, English language specification.

The TCAS example is very instructive: it shows that a formal specification can be written for a system whose complexity defied expression in natural language and that formal specification was usable by reviewers and implementors who were not experts in the specification techniques used.

3.6 HDLC

My final example illustrates the use of model checking in the area of hardware and in communication protocols—in this case a High-level Data Link Controller (HDLC) being produced by Bell Labs in Spain. The controller was initially produced using traditional hardware techniques including VHDL backed by extensive simulation. At a late stage, when the design was considered almost finished and when the builders were confident of its correctness, the formal verification team at Bell Labs offered to run some additional verification checks on the design [Calero 1997]. The checks were carried out using the FormalCheck model checking tool [DePalma 1996]. Very quickly, an error was detected that had eluded all the hours of simulation to which the design had been subjected. At best the error would have reduced throughput, more likely it would have caused lost transmissions. The model checking also helped propose a correction and this correction was itself validated using FormalCheck. Clearly much nugatory effort was avoided and model checking now forms part of the standard design process at the site concerned.

Incidentally, this is far from being an isolated example of the commitment to formal verification of hardware designs using model checking. Intel, for example, are also committed users. See for example [Schubert 2003] whose paper includes the significant words: "The principal objective of this program has been to prove design correctness rather than hunt for bugs".

4 But *Everyone* Uses Formal Methods!

The above examples show that the adoption of formal methods, in a variety of forms, is highly cost-effective and can deliver a better quality product at a lower cost (and probably to the greater satisfaction of its creators). Why then has industry not adopted such rigour with the same enthusiasm that it shows for objects, UML and automatic code generation? Despite the positive evidence we are constantly told that *no one* is using formal methods now.

Well actually, *everyone* uses formal methods, or a least a formal notation, as part of their development process; this comes as a surprise to many software engineers but is nevertheless true. The end product of any software development process, however chaotic, is a mathematically rigorous and precise description of some behaviour. That precise description is machine code and its precise meaning is defined by the target processor which provides operational semantics for the language used. The debate is not therefore *whether* to use formal notations and formal methods but *when* to use them. In the worst case scenario, we achieve precision (another word for formality) only when it is too late to help us. We have a formal specification of our system but one that can only be *animated* rather than *reasoned about*. We are forced to animate it, by testing, with all the disadvantages that brings, because we have no other choice.

By contrast, earlier adoption of more formal notations has considerable merit. We can start with specifications: the process of specification is finding a precise way of recording some desired behaviour or property. We routinely regard the end product of this process, the specification document, as being the most important aspect of specification; however, the intangible benefits that accrue *during its production* are at least as valuable. It is during the specification process that we can discover the ambiguities, inconsistencies, lack of completeness and other flaws that would eventually emerge as bugs. The more rigorous our approach to specification the more quickly and more obviously these problems emerge. As Hall puts it "It is hard to fudge a decision when writing formal specifications, so if there are errors or ambiguities in your thinking they will be mercilessly revealed: You will find you cannot write a coherent specification or that, when you present the specification to the users, they will quickly tell you that you have got it wrong. Better now than when all the programming money has been spent!" [Hall 1990].

We must therefore seek to understand why more formal approaches to software development have not generated the unstoppable momentum of UML or visual programming and, from this, find ways of injecting formality into the earlier stages of the software lifecycle where it will do most good.

5 The Situation Today

I find it useful to draw analogies with other engineering disciplines, especially aeronautical engineering which was my original profession. By the time that aeronautical engineering had advanced beyond the craft stage it had acquired a mathematical basis. For example, reasonable predictions of stress in components could be made, but only by very highly skilled engineers working very hard with tools like slide rules (that seem very primitive to us now). Today, stress calculations can be made more easily by less senior engineers using more powerful tools. The crucial thing here is that the mathematics has not been abandoned as too hard but encapsulated in a more accessible and productive form.

I believe there is a strong analogy with formal methods but with a rather less satisfactory outcome. Using the early methods such as VDM, B or Z, a highly skilled engineer (akin to our senior stress man) could make a very rigorous prediction of the behaviour of a software component. Unfortunately, because this was perceived as hard, the response was not, as in the aeronautical world, to encapsulate the mathematics but to give up doing the calculations at all. To stretch the analogy, we might liken UML to an engineer's sketch pad. We can scribble an approximation of what we want to build but we can't analyse or measure it. We are reduced to building it and then testing it to see if (or more usually, where) it breaks. This is well short of the aeronautical equivalent where we may well still test a component but in the expectation that it will not break because our calculations tell us it won't. Citing difficulty as the reason for the failure to adopt a more rigorous approach to software is especially galling when you consider how much simpler is the mathematics of, say, formal specification than that of stress or of compressible aerodynamic flows!

(In case anyone thinks this passing criticism of UML is unfair, let me quote Jos Warmer of Klasse Objecten in the Netherlands: "In many cases, people simply have their own interpretation, which is implicitly known by other people in their team, project or department. The environment and background of the reader/writer of the model determines its meaning. The consequence of this is that UML is a standard notation, but without a standard meaning."[Warmer 2003]).

5.1 Industrial Need

The failure to attempt a more rigorous approach to software development would not matter if the performance of our industry was wholly to be admired. Regrettably that is not the case. On any objective set of measures our industry paints a story of failure. Cancelled systems, late delivery, poor performance and cost overruns are the norm. Martyn Thomas [Thomas 2003] quotes figures from the Standish Group and from the BCS that show:

The Chaos Report 1995	
Projects cancelled before delivery	31%
Projects late or over budget or which deliver greatly reduced functionality	53%
Projects on time and budget	16%
Mean time overrun of projects	190%
Mean cost overrun of projects	222%
Mean functionality of intended system actually delivered	60%
BCS Review 2001	
Success rate of 1027 projects	12.7%
Success rate of 500 *development* projects	0.6%

hardly a glowing testimony to a thriving industry!

The expectation of failure has become so entrenched that it has corroded the entire basis on which contracts are offered and won. Those asking for a system to be built ask for more than they either need or expect to get. Those bidding for the work offer an unrealistically low price in the expectation that they will be able to deliver less than the contract requires or that they will be able to force the price up when there is an inevitable requirements change. The entire process is dishonest, leaves both parties dissatisfied, further lowers future expectations and discriminates against those professional organizations that are genuinely able to quote for and deliver the requested system. In the end I think this cycle of dissatisfaction actively reduces the amount of work available and may even be part of the cause for the downturn in the software business; after all, given the high expectation of failure, why would a potential customer seek to have anything built unless it was unavoidable? I suspect many systems that could be replaced or updated to the benefit of their owners soldier on precisely because of a lack of confidence that the replacement would be better or even work at all.

5.2 The State of the Art, or The State of the Practice?

What has the above told us?

- Our industry does not have a good track record for delivering dependable systems at predictable cost;

- our industry has largely rejected the adoption of more formal methods; and
- formal methods, when tried, have had an overwhelmingly positive effect on dependability and cost.

Put in that rather stark form it does suggest that formality has been unreasonably neglected and that it should be revisited. Clearly much of what is trumpeted as "state of the art" is actually no more than "state of the practice". We are not failing because of the inadequacies of computer science or because of the lack of applicable techniques, we are failing because we are not using existing methods of proven utility. As Edsger Dijkstra put it so elegantly in 1973: "Real-life problems are those that remain after you have systematically failed to apply all the known solutions".

6 The Future

So how can we move from the morass of sexy but semantic-free languages, pointless but pervasive processes and multiplicities of meaningless metrics onto some logically firmer ground? I think there are several possible routes.

6.1 Traditional Formal Methods

Traditional Formal Methods as typified by Z and VDM continue to be used and research in this area continues. Currently there is interest in the modelling of systems that exhibit a mixture of discrete and continuous behaviour. Another hot topic is the construction of notations that encompass both model-based and process-based behaviour.

The main problem with traditional Formal Methods remains one of perception and prejudice. I have had some interesting conversations with potential clients which have been proceeding very well until I have said something like: "we recommend constructing a formal model of ..." at which point the aghast client manages to combine all of Hall's seven myths into a single sentence that usually finishes with "couldn't we use UML?". In a similar vein, I know of a particularly savvy organization that produced a Z specification for a system but found that most of the implementors they offered it to wanted extra money because of the difficulty of dealing with this unusual artefact; personally I cannot think of any easier or better start to a project than to have a customer who knows exactly what they want and who can express it precisely!

The MULTOS CA project tells us that non-specialists can be readily helped to understand formal specifications and that the precision of such a specification brings contracting and commercial benefits as well as the more obvious technical ones. The production of a formal specification also potentially increases the power of a system procurer since once such a specification exists it should be possible to get the system constructed by one of several developers. By dividing a new project up into separate requirements capture, specification and construction phases risk can often be reduced. Non-formal approaches make it much more likely that the entire process has to be let to a single organization as a single contract and removes this more incremental option. For this reason alone, system procurers should not be afraid of bidders who suggest using a formal approach.

Unlike a lot of heavyweight, tool-centric approaches, the adoption of formal methods does not have to be done in a single big bang. Often it can be integrated into an existing process. It can be used to specify the most critical part of a system even if it is not adopted for everything. My experience is that even sketching out a few key safety invariants in a rigorous manner can bring useful benefits.

6.2 Formality by Stealth

This is the area that I think offers the greatest promise. It continues the aeronautical engineering analogy offered earlier by seeking to encapsulate mathematical rigour in a user-friendly wrapper. When a mechanical engineer uses a CAD program to perform a stress calculation on a model the tool will use rigorous mathematical techniques such as finite element analysis to produce the result which it will then display in a pleasant range of colours. Similarly, an aerodynamicist may explore some new wing shape using a computational fluid dynamics system. Again, he will get a visualization of fluid velocities and pressures but without direct visibility of the families of partial differential equations that the system is solving to produce them.

The growing popularity of model-based design notations such as UML and their associated tools provides an opportunity (and a risk) here. If we can develop such tools so that they become analagous to the finite element analysis and computation fluid dynamic tools mentioned earlier then we can, potentially, move formality to an earlier stage of the development lifecycle. One trend that might drive things in this direction is automatic code generation. Since machine code is formal it follows that any attempt to generate code from a diagram imposes a semantic meaning on that diagram. Unfortunately, if there is an ambiguous source language between the diagram and the machine code then the connection is rather tenuous and the imposed semantic meaning rather weak. In this case we are reduced to generating code, testing it to see what it does and tweaking the diagram if it isn't quite what we want. The situation is a little different, however, when the generated source code is unambiguous (see [Amey 2002] for a discussion of ambiguity in programming languages). Here the implied meaning of the diagram is directly deducible from the generated source code and so the diagramming language suddenly stops being semantic free. We have seen this effect with a number of tool vendors who have sought to generate SPARK from their design tools: the immediate effect is to force them to be much more precise about what their various diagrams actually mean. (This is exactly the same effect experienced on the Lockheed C130J where coders found themselves unable to express ambiguous specifications). Potentially then, we can find a sound semantics for, say, UML and then apply formally-based analysis to a model expressed in it. That is the hope, and the UML2 initiative, which has raised the emphasis on strong semantics, is a promising sign. There remains, however, the danger is that we won't even try to exploit stronger semantics for the UML. Instead we will tolerate the vagueness of the diagramming notations used on the grounds that code generation will let us get to test quickly and that testing will reveal any problems; precisely the kind of late error detection we need to escape from if we are to improve the state of the industry.

To avoid the negative outcome, users of code generation tools should challenge their vendors to explain exactly what diagram-to-code mappings are used and exactly what analysis can be performed at the model level.

6.3 Lightweight Formal Methods

The idea of lightweight formal methods was presented by Daniel Jackson at Formal Methods Europe 2001 although the proceedings do not include a full paper on the subject. An earlier version of his ideas is available on the web [Jackson 2001].

The concept of lightweight formal methods is that we are prepared to trade some of the universality and precision of formal methods to make them easier to apply to particular common problems and in common situations. The result is that we may not be able to obtain the range and precision of results that a fully formal approach offers but that we may instead get simpler and more rapid results for particular classes of problems. Crucially Jackson is not advocating abandoning the mathematical rigour of formal methods but its selective deployment with appropriate approximations. Perhaps this is analagous to the simplification that can be made to the analysis of fluid flows at low speeds where we can ignore the effects of compressibility. Our analysis is no longer universal or exact but it is much simpler and still immensely useful for a common class of problem.

Lightweight formal methods overlap with formality by stealth. For example, the Polyspace Verifier tool uses a mathematical technique called abstract interpretation to produce an approximate model of part of the behaviour of a computer program. The approximate model can produce useful results in the identification of potential run time errors. The underlying mathematics places this tool in a different class from the purely heuristic approach of, say, lint.

I have to admit to a slight scepticism with Jackson's ideas. The mathematics of computer software is so much simpler than that of aerodynamic flows that the simplifications don't always seem necessary (although simplifications may greatly speed up some analyses and make them feasible in new situations). Furthermore, for critical systems the failure to detect an error because it falls outside the approximate model being used may be unacceptable. There is for example a clear difference between *proof of the absence of run-time errors* using SPARK (a fully-formal approach) and the static detection of *some run time errors* using Polyspace.

Perhaps lightweight formal methods have most to offer in non-critical systems where an easily adopted approach that found many (but not all) common kinds of error would have a very significant effect on the overall quality of software. For critical systems, their main benefit might be in providing an initial "fast filter" that catches common problems leaving more rigorous processes to eliminate the subtle problems that this first pass misses.

7 Conclusions

Rumours of the death of formal methods are much exaggerated. The methods, in their traditional form, are still being used by the more enlightened practitioners in our industry. In particular, model checking has become an almost unremarked standard process in the design of computing hardware and in the area of communication protocols. More significantly, formality by stealth, the encapsulation of rigour in a less threatening wrapper, has started to make itself felt. Users of SPARK, for example, are employing an unambiguous language with formally-defined semantics supported by tools performing rigorous and exact analyses based on sound mathematical principles. But if you asked some of the SPARK users who are routinely using proof techniques to show their code is free from all predefined exceptions, whether they used formal methods they would probably say "no".

Business thrives on precision. We expect our contracts to be accurate and subject to a single interpretation and we employ expensive specialists to ensure that this is so. We even use standardized forms of formal address in our correspondence: "Dear Sir, yours faithfully" defines a form of interface and it is perhaps in component interfaces that formality has most to offer most quickly as we move towards the construction of systems from off-the-shelf components.

It is clear that the software industry should require the same precision and rigour in the definition and construction of its primary product as it does in any of its other activities. Don't worry about whether something carries the label "Formal Method" but *do* worry, a lot, about whether it is formal in the sense of being: *precise, rigorous, exact, amenable to reasoning;* or *susceptible to analysis*. If your suppliers cannot satisfy you that their methods, however fashionable, meet these criteria then they are falling short of currently achievable engineering standards; they are being unprofessional. If they claim their informally produced software is suitable for use in a highly-critical system, then they are being dishonest as well as unprofessional because such a claim cannot be sustained by dynamic test evidence alone [Butler 1993][Littlewood 1993]. If you, the reader, are a software supplier how well do you pass these tests?

The real benefit of a more formal approach is that it changes the software development mindset from one of *construct and debug* to the more sensible *correctness by construction*. Too often techniques such as static analysis are just seen as new and different ways of finding bugs; only formal methods—and tools based on formal methods—offer a route to avoiding the bugs in the first place.

References

[Amey 2002] Peter Amey. *Correctness by Construction: Better Can Also Be Cheaper.* CrossTalk Magazine, March 2002.

[Barnes 2003] John Barnes. *High Integrity Software - the SPARK Approach to Safety and Security.* Addison Wesley Longman, ISBN 0-321-13616-0.

[Butler 1993] Butler, Ricky W.; and Finelli, George B. *The Infeasibility of Quantifying the Reliability of Life-Critical Real-Time Software.* IEEE Transactions on Software Engineering, vol. 19, no. 1, Jan. 1993, pp 3-12.

[Calero 1997] J Calero, C Roman and G D Palma. *A practical design case using formal verification.* In Proceedings of Design-SuperCon97.

[DePalma 1996] G. DePalma and A. Glaser 1996. *Formal verification augments simulation.* Electrical Engineering Times 56.

[Hall 1990] Anthony Hall. *Seven Myths of Formal Methods.* IEEE Software, September 1990, pp 11-19.

[Hall 1996] Anthony Hall. *Using formal methods to develop an ATC information system.* IEEE Software 13(2): pp 66-76, 1996.

[Hall 2002a] Anthony Hall. *Correctness by Construction: Integrating Formality into a Commercial Development Process.* FME 2002: Formal Methods - Getting IT Right, LNCS 2391, Springer Verlag, pp 224-233.

[Hall 2002b] Anthony Hall and Roderick Chapman. *Correctness by Construction: Developing a Commercial Secure System.* IEEE Software January/February 2002, pp 18-25.

[Heimdahl 1996] Heimdahl M. and Leveson N. *Completeness and consistency in hierachical state-based requirements.* IEEE Transactions on Software Engineering SE-22, June 1996, pp 363-377.

[Hodges 1992] Quoted in: Andrew Hodges. *Alan Turing: The Enigma.* Vintage, Random House, London, ISBN 0-09-911641-3 or Walker & Co., New York, ISBN 0-802-77580-2.

[ITSEC 1991] Information Technology Security Evaluation Criteria (ITSEC). Provision Harmonised Criteria, Version 1.2, June 1991.

[Jackson 2001] See `sdg.lcs.mit.edu/~dnj/pubs/ieee96-roundtable.html`

[Keller 1993] Keller. T.W. *Achieving error-free man-rated software.* 2nd international conference on Software Testing, Analysis and Review, Monterey, California, 1993.

[King 2000] Steve King, Jonathan Hammond, Rod Chapman and Andy Pryor. *Is Proof More Cost Effective Than Testing?.* IEEE Transactions on Software Engineering Vol 26, No 8, August 2000, pp 675-686

[Littlewood 1993] Littlewood, Bev; and Strigini, Lorenzo. *Validation of Ultrahigh Dependability for Software-Based Systems.* CACM 36(11): 69-80 (1993)

[MULTOS] See `www.multos.com`

[RTCA 1992] RTCA-EUROCAE. *Software Considerations in Airborne Systems and Equipment Certification.* DO-178B/ED-12B. 1992.

[Schubert 2003] Tom Schubert. *'High Level Formal Verification of Next-Generation Microprocessors.* Proceedings of 40th Design Automation Conference. ACM Press June 2003.

[SPC 1993] Software Productivity Consortium. *Consortium requirements engineering guidebook.* Technical Report SPC-92060-CMC version 01.00.09.

[Thomas 2003] Martyn Thomas. *The Modest Software Engineer.* Presention at Software Reliability and Metrics Club, London on 20th May 2003.

[Warmer 2003] Jos Warmer. *The future of UML.* `http://www.klasse.nl/english/uml/uml2.pdf`

MATURE AND PRACTICAL FORMALITY

Safe Object-Oriented Software: The Verified Design-By-Contract Paradigm

David Crocker

Escher Technologies Ltd.
Aldershot, United Kingdom
dcrocker@eschertech.com

Abstract. In recent years, large sectors of the software development industry have moved from the procedural style of software development to an object-oriented style. Safety-critical software developers have largely resisted this trend because of concerns about verifiability of object-oriented systems. This paper outlines the benefits offered by object technology and considers the key features of the object-oriented approach from a user's perspective. We review the main issues affecting safety and propose a paradigm – Verified Design-by-Contract – that uses formal methods to facilitate the safe use of inheritance, polymorphism, dynamic binding and other features of the object-oriented approach. An outline of *Perfect Developer* – a tool supporting the Verified Design-by-Contract paradigm – is included.

1 Introduction

In recent years there has been a substantial move from procedural to object-oriented approaches in many sectors of the software development industry. The principal advantage of the object-oriented approach is the ease with which reusable components and application frameworks can be created.

Although the benefits of object technology have been oversold in some quarters, most studies indicate that software development companies moving to object technology have at worst maintained their previous productivity (Potok et al 1999) and at best increased it by several times (Port & McArthur 1999, Mamrak & Sinha 1999). The greatest productivity gains come from re-using components or frameworks from one project to another. This suggests that even if a company switching to object technology sees little saving in the short term, it will gain in the longer term as the opportunity for re-use arises. Our own experience is that in some application areas at least, an object-oriented design is significantly simpler and faster to implement than a procedural design, regardless of re-use; provided that the development staff are already experienced in object technology. We have also found object-oriented designs easier to extend to meet new requirements.

Despite the potential benefits, safety-critical software developers have largely avoided object-oriented methods, preferring instead to use procedural or modular approaches. However, a number of companies in safety-critical sectors are now moving to object-oriented software development, or planning such a move. This is particularly evident in the North American aerospace community. In recognition of this trend, a number of interested parties including the Federal Aviation Administration and NASA have set up the Object Oriented Technology in Aviation (OOTiA) programme to address safety and certification issues when object-oriented software is used in airborne applications.

This paper considers the reasons behind the slow uptake of object technology by the safety-critical software development community. We describe how the coupling of an existing design technique with formal verification allows the most powerful features of object technology to be safely used, even in critical applications. We have developed a toolset that employs modern automated reasoning technology to obtain a very high degree of automated proof, in order to make formal verification economic even for less critical software. The use of formal specifications also makes automatic code generation possible, eliminating coding errors and providing greater overall productivity than a non-formal approach in many cases. All of this makes it easier for developers to create safe software.

2 Features of Object-Oriented Technology

There is general agreement that the essential attributes of the object-oriented approach to software development include the following:

- Encapsulation: the process of encapsulating data and the operations pertaining to that data in a single entity such that the data cannot be publicly manipulated other than via the published operations. The template describing such an entity is called a *class* and plays the role of a type in procedural languages. Instances of a class are called *objects*.
- Abstraction: the process of hiding the unimportant details of an object from its users so that only the essential features that characterize it remain. The process of abstraction is greatly helped by encapsulation, since the details of how the data is represented inside an object can be hidden from its users.
- Inheritance: the principle of defining new classes by inheriting existing classes, adding new data and/or operations and possibly redefining existing operations. Some languages support single inheritance, while others support multiple inheritance (i.e. a class declaration may inherit more than one other class).

- Polymorphism: the principle that where some variable or similar entity is declared as being an instance of some class, then at run-time it may be permissible to substitute an instance of a different class derived from the original by inheritance.
- Dynamic binding (also known as *dynamic dispatch*): the principle that when a variable or similar entity is polymorphic and is passed as a parameter in a call to a function or procedure, the exact function or procedure called may not be statically determined but may depend at run-time on the class of which the variable is an instance. Most object-oriented languages support *single dynamic dispatch* (i.e. the choice of function or procedure called depends on at most one parameter, which is typically distinguished syntactically from the other parameters); a few support *multiple dynamic dispatch.*

Abstraction and encapsulation are highly beneficial features to have in a programming or modelling language and are likely to enhance safety. Indeed, the widely used *modular programming* approach captures both these features. Abstraction and encapsulation also facilitate formal analysis, since when analysing code that uses a class, the abstract specification of the class is sufficient to capture its significant behaviour, so that the detailed implementation of the class can be disregarded. Separately, the class can be analysed to ensure that its detail conforms to its abstract specification.

Inheritance does not in itself cause any particular difficulty for program analysis, since a class derived by inheritance could be expanded by substituting the member declarations of the inherited class(es) into the definition of the derived class. However, the combination of inheritance, polymorphism and dynamic binding is not directly amenable to traditional static analysis, which typically requires that the target of each procedure call is statically known.

One solution for engineers of safety-critical software who wish to adopt object technology is to eschew dynamic binding. This may be a reasonable way of getting started with object technology. However, dynamic binding has been found to be such a powerful and useful feature that this is surely not the best long-term approach. Better instead to seek new verification techniques that can ensure dependability even in the presence of dynamic binding.

3 An Object-Oriented Example

We have heard it claimed that object technology (and dynamic binding in particular) is not useful in most safety-critical software. While there may be some systems for which object technology has little to offer, there are many others that clearly could benefit from object technology provided that safety concerns can be addressed.

As a working example for the purposes of this paper, consider a glass-cockpit flight instrument display with the following requirements:

- A number of flight instruments (e.g. airspeed indicator, altimeter, horizontal situation indicator) are to be displayed on a single screen.
- The details of which instrument is displayed in which position should not be fixed in the software but should be easy to change. This will allow a family of systems to use the same software and also provide for a limited amount of installation-time or in-flight customisation (e.g. a choice of presentation, or a choice between two different instruments that convey the same information).
- There may be other information (textual information, alarms etc.) to be displayed on the screen.

This design outline was inspired by an example given in the OOTiA draft handbook (OOTiA 2003). To the object-oriented design engineer, there is a natural inheritance hierarchy in this description. At the root is an abstract class representing any self-contained displayed entity, which we will name *DisplayedElement*. Inheriting from this we might have a concrete class *TextElement* and another class *FlightInstrument*. Concrete classes such as *AirspeedIndicator* will be derived from *FlightInstrument*.

The complete glass-cockpit display can be represented by another class *Display* whose data comprises a collection of objects derived from *DisplayedElement*, each associated with the corresponding screen coordinates. To initialise the screen, we provide a method *drawAll* that iterates through the collection, drawing each object. This suggests a dynamically bound call to a *draw* method[1] that is separately defined for each concrete class derived from *DisplayedElement*. Note that the *draw* method in class *DisplayedElement* is declared but not defined, making it an ***abstract*** method (sometimes referred to as a ***pure virtual*** method). Likewise, the class DisplayedElement is an abstract class (meaning that it cannot be instantiated but serves only as a base for other classes to inherit). Classes derived from *DisplayedElement* will provide their own definitions of the *draw* method.

The architecture described above provides the flexibility we need in that any sort of *DisplayedElement* can occur at any position in the collection. Furthermore, if we wish to add a completely new type of flight instrument to this design, we need only define a corresponding class derived from *FlightInstrument* and provide a means to store an instance of this class in the collection.

Without dynamic binding, we would have to use a "switch" statement or similar in place of each call to *draw* so as to select the correct procedure according to the actual type of the element we wish to display. The same goes for any other operation that depends on the element type. If we add a new type of flight instrument, we need to update every one of these switch statements to handle the new type, thereby spreading the modification throughout the code instead of concentrating it in one place. Thus, the object-oriented approach makes it easier to

---

[1] Functions and procedures are referred to as *methods* in object-oriented software development

extend the system in a safe manner that leaves almost all of the existing system unaltered.

This example provides several opportunities for re-use. The application framework (comprising the class representing the collection of instruments and associated scheduling of *draw* operations) can be re-used for different displays with widely differing selections of flight instruments. The flight instrument classes could likewise be re-used with a different framework. Several flight instruments may share some common presentation details (e.g. style of frame and caption), so these features may be implemented in a common parent class (e.g. *SquareFlightInstrument*).

There is a standard language – the Unified Modeling Language (UML) – for describing graphically the relationships between classes (and many other aspects of object-oriented systems). A UML class diagram of the system described above is shown in Fig. 1.

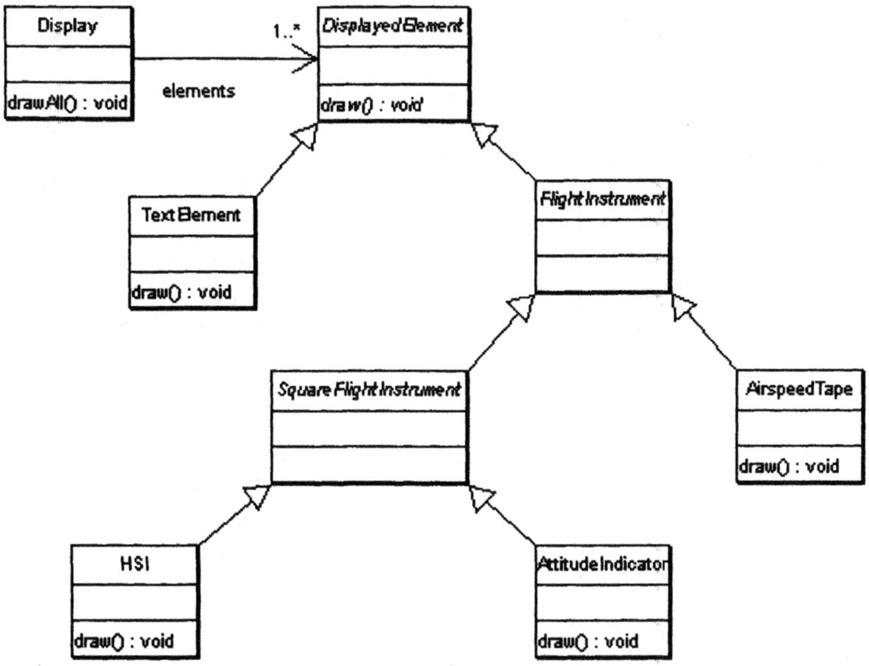

Figure 1: Class diagram for flight instrument display system

4 The Design-By-Contract Paradigm

4.1 Basic principles of Design-by-Contract

The term "Design By Contract" (DBC) appears to have been conceived by Bertrand Meyer of Interactive Software Engineering (Meyer 1988) and the term is claimed by that company as a trademark. However, the principles of DBC go back to Floyd-Hoare Logic (Hoare 1969), the essence of which can be summarised as follows.

A program statement ***S*** exists to achieve some desired *postcondition* ***R*** after its execution (where ***R*** is a predicate over the program state; in other words, ***R*** is a mathematical description of the state that the programmer intends after ***S*** is executed). Typically, the statement ***S*** will only accomplish the state ***R*** if some *precondition* ***P*** is satisfied before ***S*** is executed (***P*** is another predicate over the program state). Given ***P***, ***R*** and ***S***, then in order to be certain that ***R*** will be satisfied after executing ***S***, we need to be sure of two things:

1. Provided ***P*** is initially satisfied, executing the program fragment ***S*** will terminate in a state satisfying ***R***; and
2. ***P*** is always satisfied immediately before ***S*** is executed.

In Design By Contract, this principle is applied to the case where ***S*** is a call to a method. For each method, its precondition ***P*** and its postcondition ***R*** are documented. The correctness requirements are expressed in the form of a contract between the method and its callers, like this:

- The method promises that provided it is called in a state satisfying its precondition ***P***, then it will return in a state satisfying its postcondition ***R***.
- All callers of the method promise to satisfy ***P*** at the point of call; in return, they are entitled to assume that the call will complete and that ***R*** is satisfied on return.

Returning to our example, let's look at how DBC may be applied to the *draw* method of class *DisplayedElement* and its descendants. We assume that the parameters to *draw* include the coordinates of a rectangular portion of the screen in which the element is to be displayed. The contract might look like this:

- The caller of *draw* promises that the size of the rectangle passed to *draw* is at least the minimum needed to display the element;
- The implementation of *draw* promises that on return, the element is displayed in that rectangle and the remainder of the screen is unchanged.

It is frequently the case that the instance variables of a class should satisfy some property at all times. In our example, we might wish that in the *Display* class, the rectangles associated with the elements to be included on the display never overlap and are always wholly contained within the visible area. Such a property would logically be part of the postcondition of every constructor[2] for the class and part of both the precondition and the postcondition of every method of the class. Rather than explicitly state the property in all these preconditions and postconditions, it is simpler and clearer to state it as a *class invariant* instead.

4.2 Design-by-Contract with Dynamic Binding

The Design-by Contract paradigm can also be applied to dynamically bound method calls. We will confine ourselves to the case of single dynamic dispatch.

The general situation is as follows. The program code contains a call to a ***nominal target*** (e.g. method *draw* of the abstract class *DisplayedElement*); but this is interpreted at run-time as a call to some ***actual target*** that depends on the run-time type of the object concerned (e.g. if the object concerned has run-time type *AttitudeIndicator*, the actual target will be the version of *draw* defined in that class). Recall that the correctness of a program segment involving a method call depends on the following conditions:

1. The caller satisfies the precondition of the called method.
2. The method guarantees to satisfy its declared postcondition, provided that its precondition was satisfied.
3. On return, the caller may assume that the postcondition of the method holds.

If the method is dynamically bound, we have a potential difficulty with conditions 1 and 3 because the actual target method is not statically known (and hence we cannot determine its precondition and postcondition). Indeed, one of the features of object-oriented development is that we can add new classes (such as new classes derived from *FlightInstrument*) and that the old client code (for example the *drawAll* method of class *Display*) will work with them unaltered.

The solution is for the caller to refer to the contract of the nominal target instead of the (unknown) actual target. Our correctness conditions become:

1. The caller satisfies the precondition of the nominal target.
2. The actual target guarantees to satisfy its declared postcondition, provided that its precondition was satisfied.
3. On return, the caller may assume that the postcondition of the nominal target holds.

---

[2] A procedure whose purpose is to create and initialise objects of a class is called a *constructor*

To these we need to add:

4. Satisfaction of the precondition of the nominal target is sufficient to ensure satisfaction of the precondition of the actual target.
5. Satisfaction of the postcondition of the actual target is sufficient to ensure satisfaction of the postcondition of the nominal target.

We have added conditions 4 and 5 to link the contract that is actually satisfied with the contract that the caller assumes. We now consider how these conditions may be guaranteed.

The simple case is for each actual target to have the same contract as the corresponding nominal target. For example, suppose the definition of *draw* in class *AttitudeIndicator* inherits the contract given at the declaration of *draw* in its ancestor *DisplayedElement*. Any call whose nominal target is *draw* in class *DisplayedElement*, and which is correct with respect to the contract declared for *draw* in that class, will be correct if the actual target is *draw* in *AttitudeIndicator*.

However, it is also permissible to define a new contract for *draw* in class *AttitudeIndicator*, provided that the new contract conforms to the original. The conformance required is that the new contract may assume no more than the old, and it must promise no less. In other words:

- The overriding method (e.g. *draw* in *AttitudeIndicator*) may have a weaker precondition than the overridden one (*draw* in *DisplayedElement*), but not a stronger one (i.e. the original precondition implies the new one);
- The overriding method may have a stronger postcondition than the overridden one, but not a weaker one (i.e. the new postcondition implies the original one).

In summary, an overriding method may weaken the precondition and/or strengthen the postcondition of the overridden method.

5 Informal and Semi-Formal Use of Design-By-Contract

Design-By-Contract can be implemented in various ways, ranging from informal methods that rely on the developer to ensure correctness, to rigorous ways that are amenable to automated analysis.

5.1 Implementing DBC with Comments

The contract of a method can be documented in the form of comments. This approach may be used with any programming language. For example, a C++ declaration of the *draw* method might appear like this:

```
// File DisplayedElement.hpp
#include "Screen.hpp"

class DisplayedElement {
public:
  // Display the element in the given rectangle
  virtual void draw(Screen &s, Rectangle r)
  //pre r.height >= minHeight,
  //    r.width >= minWidth;
  //post instrument is displayed in the rectangle,
  //     rest of the display is unchanged;
    = 0;
}
```

Although contracts expressed as comments are better than nothing, it is difficult to ensure that the contracts expressed are complete and are satisfied by both parties. Design-by-Contract is much more powerful if contracts can be verified in some way.

5.2 Annotated Development with Run-Time Checks

A few programming languages and tools support notations for expressing contracts. Examples of such notations include Eiffel (Meyer 1992) (in which contracts are part of the language itself) and iContract (Kramer 1998) (an extension of Java in which specially-formed comments are used to express contracts). Typically, the tools for such languages provide the facility for generating run-time checks of preconditions, postconditions and class invariants. Nevertheless, it is still up to the software developer to ensure that the contracts are completely expressed and that the callers of a method do not rely on behaviour that is not expressed in the contract. Likewise, the developer must ensure that where preconditions and/or postconditions are redefined in an overriding method definition, the new contract conforms to the inherited contract.

Furthermore, even if run-time checks are enabled during testing, there remains the possibility that testing has not exercised all possible targets at every point of call, or that broken contracts occur in rare cases that have been missed due to insufficient coverage. Some contracts (e.g. those involving quantification over all possible values of a type) are either impossible or too expensive to check at run-time.

5.3 Annotated Development with Extended Static Analysis

Another way of using contract annotations is to perform extended static analysis (often involving term rewriting or theorem proving techniques) in order to attempt to prove that the code satisfies the specifications. The most commercially successful example of this approach is Spark Ada (Barnes 1997). Its success results from starting with a subset of Ada that avoids hard-to-verify elements such as pointers.

When this approach is applied to object-oriented languages such as C++ and Java, the following difficulties have to be confronted:

- The widespread use of pointer and reference types makes it impossible in practice to perform full static analysis except on small snippets of code because of the potential for aliasing. It is not possible to avoid reference types in Java, and it is only possible to avoid pointers and references in C++ if polymorphism is not used. In order to perform useful analysis of larger sections of code, it is necessary to make sweeping assumptions to limit the extent of aliasing. While these assumptions may frequently hold, this approach cannot be justified in safety-critical work.
- Traditional programming languages are not designed to be verifiable and have features that make verification difficult unless additional information is provided. For example, object-oriented languages typically allow a variable or parameter of any non-primitive type to have a null value. When contracts are added, it becomes necessary to include a great many preconditions, postconditions and class invariants stating that certain entities are not null. If the developer forgets to add these, the correctness conditions become unprovable.
- Complex data structures are often used to store data that is conceptually simple. For example, a tree structure may be used to store a set of records, and additional index structures may be added to enable fast searching on multiple keys. To the clients of the class that maintains this data, the internal structure is irrelevant and the operations are much better specified in terms of a simpler abstract model. Therefore the programming language needs to be supplemented not only by a means of expressing contracts but also a means of declaring an abstract data model and its relationship to implementation data.
- Programming languages do not have sufficiently powerful expression syntax to express many contracts. In particular, quantification and associated concepts from first-order predicate calculus need to be expressible; so the expression sub-language needs to be extended. This may lead to confusion in the mind of the user, because there are different expression sub-languages depending on whether the context is specification or code.
- Integer arithmetic in C++ and Java is subject to wrap-around when the result is too large to represent. This is incompatible with the interpretation that is generally required in specifications (Chalin 2003). Annotated development systems either ignore this problem or provide a context-sensitive interpretation of expressions.

The state-of-the-art in object-oriented annotated development is represented by tools such as ESC/Java (Flanagan et al 2002) which is based around annotated Java. We note that the documentation for ESC/Java states that the system is deliberately unsound in some respects – because the price for soundness in the context of Java would be a great reduction in the practical usefulness of the tool. Nevertheless, these tools represent a substantial achievement and are capable of detecting many programming errors. Where there is a requirement to assess the correctness of Java code in a safety-related project, we consider that the use of such tools would be very worthwhile as they are likely to discover many of the coding errors. What they cannot do (except in simple cases) is guarantee to find *all* of the design and coding errors (or prove the absence of any), or prove more generally that the program fulfils the requirements.

A further difficulty (affecting all forms of static analysis) is that the degree to which programs containing loops can be verified is severely restricted unless precise and complete loop invariants are available (since the program state following a loop cannot be computed without one). However, determining loop invariants is tedious and often very difficult. In order to achieve the twin goals of verifiability and productivity, the number of hand-written loops needs to be drastically reduced.

We also consider that a code-centric notation with optional annotations is far from ideal. Developers will be tempted to write the code first and add the specification annotations later. This is likely to lead to incomplete specifications and hard-to-verify code. Instead, specifications should be compulsory and central to the notation; code should be optional and subservient to specifications.

6 Verified Design-By-Contract

The limitations of informal and semi-formal implementations of Design-by-Contract are avoided if all contracts are formally verified without making assumptions that cannot be justified or sacrificing soundness in other ways. We refer to this approach as Verified Design-by-Contract.

The "Escher" project was conceived with the goal of developing a toolset to support Verified DBC with close to 100% automated verification. The system is intended for use in applications at all safety integrity levels. We now present an outline of the toolset.

6.1 Principles and Notation

In pursuit of our goal we adopted the following principles:

- Code should serve only to implement a corresponding specification;
- The notation should support specifications based on abstract data models with refinement to implementation models;
- The notation should be designed to facilitate automated verification, avoiding the problems of notations based on programming languages.

We also felt that the notation should avoid mathematical symbols that are not familiar to ordinary software developers, since many developers are put off by the highly mathematical notations of some formal languages.

These principles were embodied in the *Escher Tool*, which has been commercially released as the product *Perfect Developer*. The tool is based around a notation designed for the expression of functional requirements, specifications (of which contracts are a part) and implementation code.

Returning to our example, a declaration of the *display* method in the notation of *Perfect Developer* might look like this:

```
// File DisplayedElement.pd
import "Screen.pd";

class DisplayedElement ^=
interface

  // Display the element in the given rectangle
  deferred schema draw(s!: Screen, r: Rectangle)
    pre r.height >= minHeight,
        r.width >= minWidth
    assert isDisplayedOn(s', r),
        s'.isSameOutsideRectangle(s, r);

  deferred ghost function
              isDisplayedOn(s: Screen, r: Rectangle): bool;
end;
```

We refer to an inherited postcondition as a *postassertion* because it is necessarily incomplete; hence the 'postcondition' part of the contract of *display* is introduced by the keyword **assert**.

A 'ghost' function *isDisplayedOn* has been declared in order to properly define the first part of postassertion (i.e. that on return, the element is displayed on the screen in the specified window). This function will be defined in derived classes such that it returns **true** if the window contains exactly the displayed element (but without being concerned with the details of how it is drawn, as such detail belongs in the *draw* method). Since it has been declared **ghost**, no code will be generated for it; its declaration exists solely to facilitate specification and verification.

The second part of the postassertion expresses the requirement that calling *draw* changes no part of the screen outside the given rectangle. So we are able to specify (and verify formally) not only that *draw* correctly draws the element in the rectangle, but also that it does not corrupt any other part of the screen. The method *isSameOutsideRectangle* of class *Screen* will be another ghost function.

6.2 Verification Conditions

In common with most other formal method tools for software development, *Perfect Developer* performs type checking on the input text and generates *verification conditions* (also known as *proof obligations*). Each verification condition is a mathematical statement, and for a correct program, all the verification conditions will be true theorems. The tool is designed to ensure that, apart from a small number of documented limitations, the converse is also true: that is, if a program's verification conditions are all true theorems, the program correctly implements its specification (subject, of course, to the availability of sufficient resources and to the correct behaviour of the hardware on which it is run, the compiler and linker used to process the generated code, and the tool itself).

Verification conditions are generated to express 47 separate aspects of correctness, including the following:

- Every method precondition is satisfied at each point of call;
- Every constructor and procedure satisfies its postcondition and postassertions;
- Every function delivers its declared result value;
- When one method overrides another and declares a new contract, the new contract respects the old;
- Class invariants are established by all constructors and preserved by all methods;
- Loop invariants are established and preserved;
- Loops terminate after a finite number of iterations;
- Assertions embedded within an implementation are satisfied;
- Behavioural properties specified by the user are satisfied;
- Explicit type conversions always succeed.

By providing a mechanism to express expected behaviour, we make it possible to prove that the program satisfies safety properties and other functional requirements.

When generating the verification conditions for code, the tool computes the program state forwards from the start of each method. Initially, the known program state comprises the method precondition, the class invariant, and any declared type constraints. At any point where a verification condition is required (e.g. a method call, an assertion, or the end of the method), it generates the theorem:

$$\textit{current state} \Longrightarrow \textit{required condition}$$

where *current state* is the accumulated program state and *required condition* is the expression that should hold at that point (e.g. the precondition of a called method, or the expression asserted, or the postassertion if we are at the end of a method body).

6.3 Proving the Verification Conditions

In order to maintain high productivity, we use a fully automatic (i.e. non-interactive) theorem prover to process the verification conditions. We decided on an automatic prover because commercial software development organizations typically have neither the time nor the skilled staff needed to develop mathematical proofs, even with computer assistance.

The prover uses a combination of conditional term rewriting and a first-order theorem prover based on a modified Rasiowa-Sikorski deduction system. This combination was chosen because first-order reasoning is easier to automate than higher-order reasoning. Although some features of the notation (such as dynamic binding) cannot be expressed in first-order logic, the instances where higher-order reasoning is needed are infrequent and conform to standard patterns, so they can be handled by term rewriting.

The logic underlying the verification conditions is a logic of partial functions (because of the presence of functions with preconditions). However, it is possible to use a classical 2-valued logic in most parts of the prover, by ensuring that for any term involving partial functions, either the preconditions have been shown to hold, or there is another verification condition stating that they do so.

6.4 Reporting Successful and Failed Proofs

Automated theorem provers typically generate proofs that are hard for humans to follow. Therefore, *Perfect Developer* transforms successful proofs into a hierarchical format designed for human consumption, allowing them to be inspected or checked if required.

Failed proofs typically indicate errors. Our goal is to provide the developer with sufficient information to identify the cause of the error. This has proved to be a difficult task; nevertheless we have been moderately successful.

6.5 Developing Code from Specifications

While it is certainly possible to use Verified Design-by-Contract during the specification and design phases only, productivity can be increased by using automatic or semi-automatic code generation. Also, we have already mentioned that the correct design of loop invariants is a difficult task. This burden on the developer can be reduced if most loops can be generated automatically from specifications.

Our toolset therefore supports refinement of specifications to code (still within the same notation) not just manually but also (in many cases) automatically. Verification conditions are generated to ensure that manual refinements precisely conform to the specification. The code is then automatically refined to a slightly lower-level notation internally before being translated to a standard programming language.

6.6 Results

Perfect Developer has been used by our own organization and by others for a variety of applications. Metrics relating to three very different applications are given in Table 1. The applications illustrated are the *Perfect Developer* compiler/verifier itself, a terminal emulator, and a substantial subsystem of government information system that was originally specified using the CREATIV toolset (Warren & Oldman 2003).

	Compiler/ verifier	**Terminal emulator**	**Government IT system**
Perfect source lines[3]	114720	3192	13486
Generated C++ lines[4]	229367	6752	-
Verification conditions	13144	1349	2631
Prover success rate[5]	$\geq$ 96%	$\geq$ 98.0%	$\geq$ 99.6%
Seconds/verification condition[6]	4.5	2.4	3.8

In both projects where C++ code was generated, the number of lines of generated C++ is about twice as great as the number of lines of specification and explicit refinement. This is notwithstanding that the *Perfect* source contains comments and some specification elements (e.g. preconditions and behavioural properties) that have no counterpart in the C++. We estimate that an equivalent handwritten C++ program would contain 1.5 to 2 times the number of lines of generated C++, so the developer writes only one-third to one-quarter the amount of *Perfect* text that he/she would in C++, further reducing the opportunity to introduce errors.

The number of loops appearing in the generated C++ outnumbers loops (provided by way of explicit refinements) in the source text by a factor of thirteen to one. Thus we have succeeded in relieving the developer of much of the chore of designing loop invariants. Nevertheless, we feel that further improvement is possible in this area, since many of the remaining loops conform to a common pattern.

The lower bound of the prover success rate varies from 96% to 99.6%. In the case of the compiler/verifier, the figure is nearly two years old because we have not investigated a sufficiently large sample of failed proofs for some time.

---

[3] Including comments and extra line breaks within complex expressions to enhance readability

[4] Total of header and code files; no comments; no line breaks within complex expressions except at right margin

[5] Percentage of verification conditions that we believe to be provable for which the prover produced a proof without the need for additional proof hints

[6] Average per verification condition attempted, including unsuccessful proof attempts

Significant improvements have been made to the prover since the figure of 96% was obtained and we believe that the true figure is nearer 98% now.

The use of the tool has resulted in the detection of a number of significant bugs. For example, in the compiler/verifier, a proof failure highlighted a condition in which invalid C++ could have been generated. In the case of the terminal emulator, the protocol specification was found to contain an ambiguity, despite having been in use for five years.

7 Other Issues with Object-Oriented Development

Although the safety of polymorphism with dynamic binding is usually regarded as the most serious issue arising from the use of object-oriented technology in safety-critical systems, a number of other concerns have been raised. We comment briefly on some of these here and, where applicable, the solutions we adopted in the design of *Perfect Developer*.

7.1 Traceability

Standards such as RTCA DO-178B include the recommendation (depending on criticality level) to trace all code to requirements. In the absence of dynamic binding, this is relatively straightforward to achieve. Each procedure at the outermost layer of the software is typically present to support directly a stated functional requirement. Static analysis of the program can identify all lower-level procedures that are directly or indirectly called from the outermost procedures. Thus a lattice can be generated in which every procedure is directly or indirectly linked to one or more functional requirements.

Problems arise if a particular branch of a conditional (e.g. if- or switch-statement) is never executed because its condition can never be satisfied. Such situations can be hard to identify unless formal analysis is used. The associated code will appear to be linked to a requirement but is in reality dead or deactivated.

Dynamic binding complicates the situation because when dynamic binding is present, it is generally not possible to determine statically what method is called. However, we can take an alternative approach, based on treating method postconditions as low-level requirements. In our example:

- We define a low level requirement: "Every displayable element can be displayed in a rectangle within the screen by calling its *draw* method".
- For every class in the *DisplayedElement* hierarchy, the *draw* method is implemented so as to satisfy this requirement. The details of the implementation will vary from one instrument class to another.
- At various points in the application, in support of higher-level requirements, a flight instrument will need to be displayed in a rectangle. The programmer inserts a call to *draw* at each such point, knowing that the need coincides with the low-level requirement that *draw* satisfies.

Thus we trace the high-level requirements of the application to the low-level requirements associated with called methods such as *draw* (possibly going through some statically-bound method calls on the way). Separately, for each class derived from *DisplayedElement*, we trace the implementation of *draw* to the low-level requirements defined for that method. If each low-level requirement can be traced to some high-level requirement in this way, and every piece of code can be traced to some high-level or low-level requirement, we have achieved traceability even in the presence of dynamic binding.

Problems arise if a method is never called for some class(es); for example, we might define a type of *DisplayedElement* that is never displayed. Again, this situation can only be found in the general case by formal analysis.

We note than when formal verification is performed, the proofs that requirements are met contain all the information needed to trace the requirements to the code that implements them. It is our intention to extend *Perfect Developer* to generate a trace lattice automatically from the proofs.

7.2 Worst Case Execution Timing

In real-time systems it is required that certain program segments complete within defined deadlines. Where the program includes method calls that are subject to dynamic dispatch, the execution time will depend on the methods actually called and it is therefore difficult to determine statically.

A solution is to divide up the maximum allowable execution time of a program segment into a budget for each individual method call and a remainder for other statements. This can be done in such a way that the total execution time (taking account of any loops involved) will meet the deadline, as long as no individual method call exceeds its budgeted time.

To ensure by design that each method call completes within its budget, we can include the time budget in the contract of the nominal target. The method's side of the contract now reads:

- Provided my precondition ***P*** is satisfied on entry, I promise to return within time ***T*** in a state satisfying my postcondition ***R***.

When one method declaration overrides another, the time budget of the overridden declaration is inherited by the overriding declaration by default, just like the precondition and postcondition.

We saw previously that when one method overrides another, instead of inheriting the contract as-is, it may improve on it (i.e. require less and/or deliver more). Provided we are only interested in maximum execution times (and not also in minimum execution times), the contract of an overriding method might improve on the overridden contract by promising to complete in a shorter time.

7.3 Dynamic Memory Allocation

Dynamic memory allocation is generally avoided in safety-critical software. This policy is typically justified on the grounds that memory allocation operations may fail due to insufficient memory or excessive fragmentation, and that the time taken to perform them will vary depending on the history of calls to allocate and release memory.

Object-oriented programming languages typically rely on dynamic memory allocation to allocate all objects of non-primitive types. Therefore, if safety-critical systems are constructed using object technology, it is necessary to establish policies for the safe use of dynamic memory allocation. We suggest here two such policies.

The first policy is to use dynamic memory allocation during the initialisation phase only. In our flight instrument example, we would expect that the set of all elements that might need to be displayed is known at the start. We can therefore create them all during initialisation, even if not all of them need to be displayed immediately. Other objects (such as values of type *Rectangle* in our example) can be implemented as value types, avoiding the need for dynamic memory allocation when creating them.

This policy is comparable to allocating all data statically when a procedural approach is used. It is likely to be adequate for many safety-critical systems. However, it would not suit a system that handles a varying number of objects, such as an air-traffic control system handing a varying number of aircraft.

Our second policy covers this situation by maintaining a free-list for each class that has a varying number of instances. In order to run in bounded memory, there must be a known upper bound on the number of instances of each class. We can initialise each free-list with the corresponding number of instances. Provided the upper bounds are respected (which will be typically be enforced by class invariants), there will never be a need for dynamic memory allocation other than from the free-lists. Allocating from a free-list in bounded time can be easily implemented. The free-list mechanism can be provided either by the supplier of the compiler and associated libraries (as we do with *Perfect Developer*), or (in some languages) by declaring a custom allocator for the classes concerned.

This policy is, in essence, similar to declaring a static array of objects tagged by 'in use' flags, allocating and releasing slots in the array as objects are created and destroyed.

If we have control over the standard memory allocator, we can choose not to populate the free list in advance. When allocating an object, if the corresponding free list is empty then we use the standard memory allocation mechanism instead. Provided that no memory has ever been released via the standard mechanism, allocation will not involve searching multiple free blocks and can therefore be performed quickly. We still need an upper bound on the number of objects of each type so that we can compute the maximum amount of memory needed and ensure that this amount is available.

7.4 Overloading

Object-oriented languages typically provide a mechanism for declaring multiple methods with the same name, distinguished only by the numbers and/or types of parameters. This mechanism is known as *overloading.* The compiler decides which declaration is the intended target of a method call by choosing the one whose formal parameter list best matches the actual parameters, according to some set of criteria.

Overloading can be very useful and, by itself, is not dangerous. The symbol "+" has long been used to stand not only for the addition of integers but also for the addition of real numbers. Similarly, it is quite natural to use the call "print(expression)" where we are happy to accept the default printing format, and "print(expression, format)" where we wish to be more specific.

However, we consider that the combination of overloading with automatic type conversion is dangerous because it brings the possibility that more than one method declaration may match a particular call. The choice made by the compiler may not correspond to the intention of the user, who may not have realized that the ambiguity existed. The situation is even worse if the language also allows trailing parameters to be omitted in actual parameter lists by providing default values, as in C++.

The solution we adopted is not to perform implicit type conversions (these are, in any case, undesirable in languages used for safety-critical software development). We make one exception to this rule to allow the type of a value to be automatically converted to a supertype (otherwise the notation becomes very clumsy to use). This exception raises the possibility of ambiguity, so we explicitly forbid any instance of overloading for which it is possible to construct an ambiguous call. We have not found this restriction to be onerous in practice; indeed, we find that the corresponding error message is only triggered where a mistake has been made, or the same name has been used for two unrelated operations.

7.5 Template Instantiation

Templates (known as *generics* in Ada) are a feature of many object-oriented languages. They have been found to be very useful in developing re-usable components, especially for representing collections of objects.

However, it is possible for template declarations to make assumptions about the types with which they are instantiated. This carries the risk that a template may be instantiated with types that violates these assumptions.

One way of avoiding this danger is to perform formal verification of each template separately for each type with which it is instantiated. Although simple in concept, this has the drawback that the verification process is substantially lengthened if there are many different instantiations.

The solution we adopted is to make the assumptions explicit by providing syntax for *instantiation preconditions* in the specification notation. We create a contract between a template declaration and the code that instantiates it, similar to the contract between a method and its callers. The template declaration is formally verified just once and the instantiation preconditions are assumed to hold during this process.

7.6 Reference Semantics and Aliasing

In most object-oriented languages, objects are assigned and copied by reference: that is, a pointer to the object is copied rather than the object itself. Some languages (e.g. C++) also support assignment and parameter passing by value, although polymorphism is typically not available then.

It is well known that the presence (or even the mere possibility) of multiple pointers or references to a common object causes substantial problems for static analysis and formal verification. The provision of reference semantics by default is also a source of program errors, such as the use of a normal assignment or equality operation where cloning or deep equality is needed to achieve the desired result. A classic example is where a class is declared to represent a text string. It is natural for strings to have value semantics; yet in Java and some other languages, the *String* class has reference semantics. The best that the designers of Java were able to do to ameliorate this situation was to provide two classes: an immutable *String* class (for which reference semantics are safe since no modification of the object is possible) and a mutable *StringBuffer* class.

In our example, reference semantics are unlikely to be a problem for objects of classes derived from *DisplayedElement* since we are unlikely to store references to them other than at a single place in class *Display*. However, if objects of class *Rectangle* have reference semantics, this may well be troublesome because we are likely to refer to values of type *Rectangle* in many different places.

The solution we adopted is to specify that in our notation, objects of all classes and types shall obey value semantics. Where aliasing is required, variables of reference type may be declared. We have found that in practice, reference variables are rarely needed.

7.7 Inlining

The C++ language supports the **inline** keyword and provides default inlining of methods whose definitions are included within their declarations.

Inlining makes it more difficult to verify that the object code conforms to the source code. However, should this prove to be a problem, most compilers allow inlining to be disabled.

7.8 Suitability of Mainstream Object-Oriented Programming Languages

Safety-critical developers rightly complain about the unsuitability of mainstream object-oriented programming languages for critical systems. The most widely used object-oriented programming language is C++, which inherits nearly all the problems of C and adds a few new ones such as ambiguous method calls.

We agree that use of handwritten C++ code carries risk in safety-critical systems. However, we observe that C is widely used in critical systems, usually in the form of a subset such as MISRA, conformance with which can mostly be checked statically. We consider that MISRA C could be readily extended to include a subset of C++. Ambiguous method calls could be banned, while constructs such as pointer-to-member and the more esoteric features of templates could be excluded. Although the use of such a subset is not an ideal solution, we believe that it could be safer to use than plain MISRA C due to the increased encapsulation available in C++ and the availability of better alternatives to troublesome features of C.

Although it inherits much of the syntactic idiosyncrasy of C++, Java is a somewhat safer language. Unfortunately, its lack of support for user-defined types with value semantics increases the need for dynamic memory allocation. The Microsoft language C# is in many ways similar to Java but supports value types. The provision of a garbage collector in both languages is a boon to commercial software developers but is likely to be unacceptable in real-time systems. This may be less of a problem in future as generational and concurrent garbage collectors (Jones & Lins 1996) become mainstream. We note that a real-time subset of Java has been defined (RTJ 2003).

The Ada 95 language extends the Ada 83 standard by providing support for (among other things) polymorphism and dynamic binding. However, we do not regard Ada 95 as a satisfactory language for object-oriented development. Unlike other languages, it does not syntactically distinguish the parameter on which dynamic binding depends from other parameters, which we consider likely to cause confusion. The notorious "with-ing" problem makes it impossible to construct complex object-oriented systems unless an ugly workaround is used. We understand that both issues are to be addressed in the next revision of Ada.

The ideal solution is to use a language that does not have its roots in C and is designed with correctness and type-safety in mind. The Eiffel programming language is certainly much better designed than the mainstream object-oriented languages but has not been widely adopted.

Many of these concerns are of little or no importance when complete code is generated automatically from specifications expressed in a rigorous notation. The primary requirement is to ensure that the compiler implements the semantics assumed by the code generator. This can be achieved by generating code in a language subset carefully chosen to avoid areas of undefined behaviour and complex constructs that might be troublesome for the compiler. We adopted this approach in the code generators of *Perfect Developer*. We note that when the tool is configured to generate code in C++, the generated code conforms to nearly all the MISRA C rules, even though we were unaware of MISRA when the code generators were specified. This suggests that we and the authors of MISRA C had similar ideas on which features of C should be avoided.

7.9 Unified Modeling Language (UML)

UML is the most widely used graphical notation for object-oriented analysis and design and is supported by a wide range of tools. Most tools can generate code skeletons from UML diagrams; some go further and claim to generate complete code if enough information is provided.

Despite the widespread marketing of UML tools, it appears that many object-oriented developers - perhaps the majority - manage without them. However, the use of UML may increase as more universities include UML in their computer science courses, and as open-source UML tools mature.

Concerns about UML among safety-critical software developers centre on the lack of a precise semantic definition for the language.

We consider that UML is a useful notation for displaying graphically the structure of a system and the relationships between the system, its components and its users. However, UML is not a substitute for a precise formal specification. Although UML has a formal sub-language called Object Constraint Language (OCL), it is rarely used, poorly supported by commercial tools and much less expressive than *Perfect Developer* notation.

Our toolset therefore allows UML models to be imported and will generate the corresponding skeletons; but the user must add the detailed requirements and specifications. A future version may allow the *Perfect* specifications and refinements to be embedded in the UML model itself.

8 Conclusions

Object technology undoubtedly facilitates re-use to a greater extent than previous programming paradigms. This is clear from the widespread existence and use of application frameworks and large component libraries. It is likely that without object technology, it would not have been economic to develop many of today's complex and powerful commercial applications.

Safety-critical software developers are right to be cautious in adopting new technology; but rather than dismissing object technology because it is not amenable to yesterday's verification techniques, the safety-critical community should seek new techniques to facilitate safe use of the new technology. The most important new issue arising from object technology is polymorphism with dynamic binding, which is tamed by the Design-by-Contract principle. The use of modern formal methods technology to implement Verified Design-by-Contract provides a basis for safely harnessing the power of object technology in critical systems.

References

Barnes J (1997). *High Integrity Ada: the SPARK Approach.* Addison-Wesley, England.

Chalin P (2003). Improving JML: For a Safer and More Effective Language. *FME 2003: Formal Methods* (Springer LNCS 2805): 440.

Flanagan C, Leino K.R.M, Lillibridge M, Nelson C, Saxe J and Stata R (2002). Extended static checking for Java. *Proc. PLDI, SIGPLAN Notices* 37(5): 234-245.

Hoare C.A.R (1969). An axiomatic basis for computer programming, *Communications of the ACM* 12: 576-580.

Jones R and Lins R (1996). *Garbage Collection.* Wiley, England.

Kramer R (1998). iContract - The Java(tm) Design by Contract(tm) Tool. *Technology of Object-Oriented Languages and Systems*, August 03 – 07: 295.

Mamrak A and Sinha S (1999): A case study: productivity and quality gains using an object-oriented framework. *Software - Practice and Experience* 29(6): 501-518.

Meyer B (1988). *Object-Oriented Software Construction.* Prentice Hall, England.

Meyer B (1992). *Eiffel: The Language.* Prentice Hall.

OOTiA (2003). Handbook for Object-Oriented Technology in Aviation (draft). *OOTiA Workshop Proceedings*, March 5, 2003.

Port D and McArthur M (1999). A Study of Productivity and Efficiency for Object-Oriented Methods and Languages. *APSEC 1999* (IEEE Computer Society): 128-.

Potok T, Vouk M and Rindos A (1999). Productivity Analysis of Object-Oriented Software Developed in a Commercial Environment. *Software - Practice and Experience* 29(10): 833-847.

RTJ (2003). http://www.rtj.org (30 September 2003).

Warren J.H and Oldman R.D (2003). A Rigorous Specification Technique for High Quality Software, *Proceedings of the Twelfth Safety-Critical Systems Symposium*, Springer-Verlag (London).

A Rigorous Specification Technique for High Quality Software

John H. Warren & Robin D. Oldman
Precision Design Technology Ltd.
6 Kings Grove, Maidenhead, Berkshire, England

Abstract

Too many software projects fail. One important reason, though not the only one, is the absence of a good specification. Specifications should be complete, consistent, comprehensible, and correct. Correctness can only be demonstrated if the specification is formal (so that reasoning can be supported); but the associated use of a formal language seriously reduces user comprehension, so there is a conflict between these two properties. We contend that formal methods should be used but that their use should be totally concealed and automated, so that users are unaware of the underlying formality. We have constructed a specification toolset, called CREATIV, which embodies this approach.

The use of formal methods mandates a scientific approach. One possible approach is to formalise specification knowledge as an axiomatic system. The CREATIV toolset uses a new model and a new definition of the specification process, together with an axiomatic theory to support specification knowledge. All operations in the system are provable and traceable; we have built the reasoning component of the CREATIV toolset on the basis of this theory.

We have used the toolset for specification on a range of projects. More recently, we have used it on a small number of government projects. We report on some of the advantages of its use, and offer some preliminary comments on a comparative specification exercise.

1 Overview

Too many software development projects end in failure, with only one in six projects being successful. One important reason, (possibly the most important, according to some surveys), is poor requirements analysis, definition, and specification. My purpose in presenting this paper is to present a new and rigorous approach to the preparation of requirements specifications for computer and information systems. It is a requirement of this approach, called CREATIV, that it should yield requirements specifications that are complete, consistent, correct, and comprehensible. This demands a formal and mathematical approach to the preparation of specifications, which is not especially popular with many analysts and specification writers. For this reason, all the mathematical formality and reasoning should be concealed from users and made entirely automatic. The next section of this paper (Section 2) describes why good specification is important and why this concealment is both attractive and necessary.

If specification is to be a rigorous discipline, then there must be a proper scientific approach to the subject. We suggest a new approach to specification, in which all specification knowledge is embedded within an axiomatic or rule-based theory, though this theory is normally concealed from users. This requires:

- a new but intuitively reasonable model of the specification process;
- one or more rules for reasoning about specifications (rules of inference); and
- a set of proven theorems on which new specifications can be based.

Since the essential "reasoning" part of the approach is itself an information system, it can be specified using its own notation and shown to be correct according to its own rules of reasoning. Section 3 of this paper describes the model, Section 4 the theory, and Section 5 describes the construction of the reasoning component of the toolset.

Given the model, its theory and a toolset constructed to implement the theory, we can construct better specifications (Section 6). This is achieved by an iterative process: adding data to the existing specification until it is *complete*, compiling and correcting the specification until it is *consistent*, and animating and amending the specification until it is *correct*. This approach offers several advantages, including positive user involvement and tangible progress measurement. A small example specification is described, together with some possible extensions. This section also describes, briefly, ways in which the repository content can be presented, not only for examination, review, and use as a specification but also for collaborative use with other toolsets, one of which is described. We also describe some practical experience from the preliminary use of the toolset.

2 The Importance of Good Specification

2.1 The Software Problem

We assert that too many software development projects end in failure. This harsh view could be ameliorated, from "failure" to "less successful than they might be", but in neither portrayal is it possible to avoid the issue that improvement is both necessary and important. If evidence is needed to support this position:

- The BCS Review (2001) surveyed 1027 projects of which only 130 (less than 13%) were judged to be successful. Of the 500+ development projects in the sample, only 3 (0.6%) succeeded. Of the 130 successful projects, 18.2% were maintenance projects and 79.5% were data conversion projects; only 2.3 % were development projects. *Software development is not done well.*
- The Chaos Report (1994) reported 31% of projects cancelled before delivery and 53% exceeding timescale and costing 189% of their original estimates and/or delivering greatly reduced functionality (60% on average). Only 16% of projects delivered specified functionality to time and budget.

If 13% - 16% is taken as the percentage range for successful projects, then at least five projects out of every six are not successful. This is a shameful assessment if software developers wish to be regarded as professional. What would be our opinion of surgeons when five operations in every six were not successful or of civil engineers if five bridges out of every six had to be closed for rework after opening? The fact remains that such is not the case: failing surgical and bridge building projects collect very considerable adverse publicity, in part simply because they are so rare.

There are of course very many reasons why projects depart from their intended path. Often there is no single cause, such departures having several proximate causes rather than a single one. We suggest that one contributory factor that is frequent and important is the absence of a good requirements specification. Additional data in the BCS Review (2001) confirms this; failure occurred most frequently and most importantly at the requirements definition stage and the most common and most important cause of failure was unclear objectives and requirements. It is on this requirements specification aspect of a software project that we wish to concentrate.

2.2 Principles of Good Specification

We suggest that a good specification should endeavour to attain four principal objectives: it should be:

- consistent
- complete
- comprehensible; and
- correct.

These objectives are generally not achievable without substantial computer assistance provided by some form of CASE[1] tool. It is important that any new approach and any new toolset should be judged on its ability to meet these four objectives. While there are many other objectives that a specification should attain, in this paper we concentrate only on the above four.

Consistency for Specifications

An inconsistent specification is undesirable, because it permits two (or more) potentially conflicting interpretations to be made from one specification document. Ideally, each inconsistency would be recognised, queried, and resolved; more often, the inconsistency is not recognised and different parts of the finished system simply work in different ways, with both client and builders regarding themselves as blameless. Systems with inconsistencies can also be harder to maintain.

Completeness for Specifications

An incomplete specification is undesirable, because the delivered system will be at best as incomplete as was the specification and the client will need to commission further work to enhance the delivered system so as to include the additional facilities that were omitted from the specification. Such enhancements

[1] An acronym for Computer Aided Software Engineering

can occasionally be disruptive to the underlying architecture if they represent facilities that the original specification could not have supported.

Correctness of Specifications

An incorrect specification is undesirable, because it defines a system other than the one the user wants. Many current approaches to specification can only be checked for correctness by review, which is generally inadequate. It is highly desirable that specification correctness be confirmed. This should be done in part by testing and in part by reasoning; reasoning frequently implies some form of proof. Section 2.3 provides more detail about specification correctness.

Comprehensibility of Specifications

An incomprehensible specification is undesirable, because it cannot be understood by readers; the prime purpose of a specification is to communicate the users' requirements to those who will design, build, and test the proposed system. If the specification cannot be understood, then communication with those who implement the system will not be possible. Section 2.4 provides more detail about specification comprehensibility or understanding.

2.3 Specification Correctness

It is now widely recognised that testing can reveal the presence of errors but never their absence. This means that specification correctness can never be shown by testing. This is not intended to denigrate testing; a tested specification is far better than an untested one, even though a proven one would be even better (though still not necessarily perfect).

While errors may be exposed by testing, correctness can only be shown by reasoning. It is not possible to conduct formal reasoning about informal specifications, and so specifications must be made mathematically formal at some point in their development if correctness reasoning is to be undertaken. This raises a practical difficulty: many analysts and specification writers find mathematical formality unattractive, to some extent because they may lack training and practice in the techniques involved.

Mathematical and formal techniques are essential to correctness demonstration for specifications but too few practitioners are available with these skills. One way to avoid this impasse is to encapsulate the mathematical formalisations and to automate them in such a way that the specifier is unaware that he or she is receiving support from formal methods.

2.4 Specification Comprehension

A specification should achieve an identical understanding in the mind of each of its readers.

From what has been said earlier, the use of a mathematically formal language or notation is a necessary adjunct to the use of formal techniques; without these we cannot show correctness. Encapsulating the formal methods is of little use if we have to feed them with texts written in formal languages. We are bound to use a formal notation or language internally for specification but we must use some other notation when communicating with people. We should therefore repeat the encapsulation process adopted earlier, whereby the formal language components of the toolset are encapsulated and operate with no visibility to the user. There must therefore be languages for input and output that are "user-friendly" and these must be translated to and from the formal notation adopted internally.

For input, we use a simple "form filling" dialogue in which users may complete each form in whole or part. A number of forms are provided that are used at different points in the specification process. The internal reasoning processes require only that mandatory values be provided on each form; optional values may be omitted. Partially complete forms may be recovered for later expansion or enhancement.

For output, we provide several presentation languages, such that output in these languages can be constructed automatically (by rule-based translators). Each form of presentation can then be customised to the particular needs of one group of users. For example, we generate an overview diagram and detailed tables of information as part of the specification process; but we have also produced Z texts as well as translations in other notations (SSADM dataflow and entity life history diagrams). Section 6.5 describes the use of translation to interface to another toolset.

Note that each generated text will accord exactly with the underlying specification knowledge repository because there is no manual involvement in its production. If "exact accordance" is not obtained, this can only be because of a translator error; once corrected, that error vanishes for all time and so the translation quality improves steadily rather than remaining constant.

This approach allows all the reasoning and translation processes to be fully automatic and totally concealed. Most of the benefits of formal methods can be obtained without any user having to become highly expert in their use. Melham, quoted in (Pierce, 2002), commented that "*Formal methods will never have a significant impact until they can be used by people that don't understand them.*" This concealed approach is one way of introducing and using formal methods to achieve the benefits they bring without requiring deep user understanding; it contributes substantially to the understanding of many senior and highly intelligent people in organisations who need to know and measure what is happening (and may also sign the cheque for your efforts!).

3 A Model for Specification

3.1 Introduction

Our principal objective in this paper is to describe one practical solution to these problems. Our approach to this solution has three parts:

- the delineation of a model for specification (Section 3);
- the development of a theory of specification (Section 4);
- the construction of a toolset to implement that theory (Section 5); and
- the use of that toolset in practical situations (Section 6).

We have so far identified some of the principal requirements of good specifications and briefly summarised why they are hard to achieve. To provide a solution to these problems, we require a sound formal and mathematical theory in which correctness reasoning can take place. We also require a toolset to implement this theory. We demand that the toolset that carries out reasoning in the theory be constructed according to the theory; the toolset must be demonstrably correct before its reasoning can be trusted. If the theory is sound and the toolset is trustworthy, then both can be used with confidence for better specification.

3.2 Components of a Specification Model

Abstraction is the process of paying strict attention to those aspects of a phenomenon that are important (for a given purpose) and not paying attention to (ignoring) those that are not. A *model* is one form of abstraction, and one which operates according to the scientific principles embodied in the theory underlying its construction. Specification is concerned with building models of real world systems, primarily for information processing.

The word "specification" is itself ambiguous since it can refer either to the system requirements definition itself or to the process (activity) of producing this definition. Where specification is regarded as an activity, this activity can be specified and so the activity or process of specification can itself be specified.

The specification language that we use is a formal language for system or process description. The same language can also be used to describe the process of building such descriptions. The specification (description) of the specification (process) can therefore be written in its own language; stated another way, the language is its own metalanguage (where a metalanguage is a language or a system of symbols used to discuss another language or symbolic system).

Every information system exists to provide one or more services by which it returns value to its users. Each such service can only be provided as a result of one or more activities undertaken by the system. In general, these activities will include accepting data values from the external world, computing new data values from existing ones, and transferring data from one activity to another in accordance with certain rules (of business, of logic, and of the physical universe). An information system can therefore be modelled as a collection of interacting, concurrent, and asynchronous activities. These activities form a connected network, each activity interacting with other activities in a controlled and rule-based manner.

The <Data> Component

Each system service is provided by the computation of one or more data values (a system in which no data value can be calculated is trivial). Each activity requires its own particular subset of the total system data in order to fulfil its obligations within the system. Every specification must therefore define the *data* structures required by each activity. The notion of <data> is, whether directly or indirectly, prerequisite to every other notion in a specification.

The <Function> Component

Some values to be assigned to the elements of some data structures are computed, usually by reference to other values already present. Every specification must therefore define a computational function to be optionally associated with each activity. These functions must have a mathematically precise definition. The notion of data is prerequisite to the notion of <function>, since functions accept data as input and/or return data as output.

The <Concept> Component

Within this model of the specification process, the specification of an application is composed of concepts, which are the elemental units of data and function; these are necessary to represent a logical unit of behaviour in a specification. This approach allows more precise specification, since each concept definition shows the precise data applicable in each well-defined concept and hence the precise activity of each function at that point in the specification (note that one function may be used with different data at other places within the specification). The two notions of data and function are both prerequisite to the notion of <concept>, since concepts require (mandatory) data and (optional) function references in their definition.

The <Event> Component

The activities do not proceed in perpetuity. They are quiescent until the arrival of an event initiates some activity. An event achieves this by providing relevant data to a concept; computation is performed on this data. Once this is complete, activity ceases until the arrival of the next event. Every specification must therefore define the events to which each activity will respond. The notion of concept is prerequisite to the notion of <event>, since the data structure of the one or more events that may activate one concept is defined by that concept.

The <Constraint> Component

Every information system, and therefore specification, must recognise and support the business rules that govern the application. These rules control the operation of the system ("when things happen"). Business rules, or constraints, are vitally important not only to the successful deployment of the system but also to the success of the enterprise; they must therefore be included in every information system specification. The notion of function is prerequisite to the notion of <constraint>, since an operation (function) is a necessary component of a constraint definition; the notion of concept is doubly prerequisite to the notion of <constraint>, since a constraint is a relationship between two different concepts.

Inter-Relation between Components

The diagram (Figure 1) shows the way in which these five structural components are inter-dependent. Each soft box encapsulates the complexity of one of the five components: data, function, concept, event, and constraint. Each arrow represents a possibly complex set of constraints showing the dependence of the lower component on the higher one. Within each structural component (soft box) there is further and more detailed "conceptual" structure; there are constraints between the individual concepts within each structural component shown in Figure 1 as well as constraints between concepts in different components.

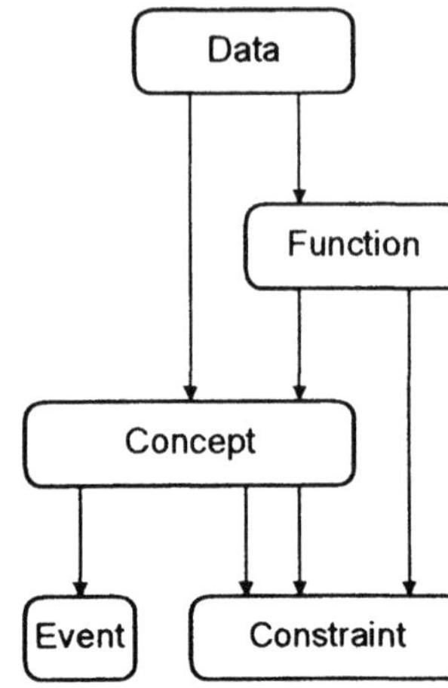

Figure 1

Constraint Satisfaction

A constraint has a formal definition, invariant with time, which expresses its precise purpose. During the system lifetime, circumstances may arise in which the constraint becomes true for certain data in the system. This will be detected during evaluation of the constraint; whenever the evaluation is true, the relevant action may be undertaken. This "constraint satisfaction" approach is an Artificial Intelligence paradigm in which problems are defined in terms of a set of constraints. The specification correctness problem can be expressed in this way: if a precise definition of the constraints defining the problem can be given, then automated reasoning techniques can be applied to identify a solution.

Constraints and Information Flow

The principle of constraint satisfaction requires that the constraint set be satisfied before any state change can take place. It is therefore necessary that the constraining information be already present in the system. This information must be stored in other concepts, since data may be stored nowhere else; these concepts must therefore stand as prerequisites to the concept being constrained and hence be logically connected to it. Constraints therefore form the interconnection between concepts. Since the constraints allow legitimate data values from prerequisite concepts to be used in the dependent concept, constraints also define data flows within the system, though only valid data can traverse the flow line. This contrasts with most dataflow diagrams, where there is little or no definition of what data traverses which flow line and under what conditions.

3.3 A Definition of Specification

What is a specification? The most frequent reply is that a specification is "a document that states *what* the proposed system should do without regard for *how* this is to be achieved". This is not a definition but is rather a description of one characteristic of a specification. In particular, it is not helpful to less experienced analysts or specification writers because it is not constructive – it gives no indication of how to achieve such a document. At least one graduate textbook on specification does not even have an index entry for the term!

An Alternative Definition & its Purpose

Given our model of specification, we offer an alternative definition: a specification is the collection of sets (data, function, concept, event, constraint) and their contents, whose members are consistent because they satisfy a particular fixed set of constraints (within and between sets). This definition identifies all the concepts needed to complete the formal model, and all the rules (constraints) that each element of each concept must satisfy. (Please note that the complete complexity of the specification model has not been defined in detail in this paper).

The purpose of this definition is that it guides the analyst and specifier in their work, by identifying all the items of information that are needful to complete the specification; that is, it is constructive. In addition, it makes evident all the rules that each entry must satisfy; these will be enforced by the compiler.

The Value of a Scientific Theory

A scientific theory is the exposition of the abstract principles of a science. A science is (a body of) knowledge that has been systematised and brought under general principles. It is these general principles that are valuable, because they provide expert guidance to those involved in the specification process. We need a specification theory because these general principles (or laws, or rules) allow us to:

- *reason* about the properties of a specification (for example, consistency)
- *predict* the consequences of a given situation (for example, by animation)
- *prove* that certain properties of a specification are present (or absent)

The advantage of a mathematically formal theory of specification is that, for any specification constructed according to the theory, we can reason about that specification: we can check consistency by rule, animate a specification to show its behaviour, and prove that desirable properties are present.

In summary, our model for specification is based on a network in which a collection of activities is represented by nodes, each of which is linked to others by constraints or business rules, represented by lines or directed arcs. Each activity (node) requires data on which to operate, may involve a computational function on that data, and is activated by one or more events. Each activity must obey the business rules (satisfy the constraints) to which it is subject. These rules or constraints are also used to generate additional information by inference. This model will operate according to scientific principles; these principles are embedded in a scientific theory. One way, though not the only way, of organising such principles, is as an axiomatic system.

4 Specification as an Axiomatic Theory

One of the major concepts in mathematics is that of an axiomatic theory. Such a theory is defined by a small number of primitive terms, one or more rules of inference, and a set of statements that are true within that theory. These statements are categorised into two disjoint sets. The first of these sets, known as axioms, contains those statements that are assumed to be true. The second of these sets, known as theorems, contains those statements that can be derived from the axioms by means of rules of inference. Euclidean geometry is probably the best known example of an axiomatic system. Each of these defining items (primitive terms, rules of inference, axioms, and theorems) will now be described.

Primitive Terms

The present version of our toolset, called CREATIV, relies upon the relational data model as a formally sound model for information storage and manipulation. It is possible to formalise this model (for example, in Z or in our own notation, and both of these have been done) but this would add very considerably to the information presented here without advancing the specification issues by any significant amount. Within this paper, we therefore propose to treat the relational data model as "primitive" and assume that relational terms such as domain, attribute, and relation can be freely used, together with the relational operators where these may be needed. A small number of software terms (such as *class* and *object*) are also used as though they were primitive.

Rule of Inference

The reasoning system, which provides the formal part of the CREATIV toolset, is written in Prolog and hence resolution theorem proving is natural. The resolution method proves a theorem by showing that its negation is unsatisfiable. Within CREATIV, there is one rule of inference, which contains five components. The rule is one rule with five components, rather than five independent rules, because all of the components must succeed in their entirety, otherwise the rule fails. This rule inserts one tuple (or "record") into the specification knowledge repository[2], and must be accepted as reasonable, in much the same way as axioms must be accepted as truths, the correctness of which is self-evident. Given that our approach controls information according to the tenets of the relational data model, and that each class is represented as a relation, the following are the five components of the rule that must be accepted. Every tuple or "record" of data values to be entered into a relation:

1. may be subject to a defined computation. Less formally, an algorithm may be invoked to create new data values in the tuple (for example, where no value but only a place holder existed previously) or to amend existing data values, though the number of attributes in the tuple must remain unaltered
2. must agree with the signature of the relation in degree and data type. Less formally, there must be exactly one data value in the tuple corresponding to each attribute in the signature of the relation and the data value must be of the same data type (or defined on the same underlying domain of values) as was the attribute
3. must be uniquely identifiable by the primary key of that relation. Less formally, there must be a set of attributes within the signature of each relation such that, for all time, it is the case that no two distinct tuples in the relation

---

[2] In practice (for increased execution speed), we may use two rules: one to *insert* and one to *remove* tuples from the repository. Since *remove* can be expressed in terms of *negation* and *insertion*, the *remove* rule is theoretically superfluous.

have the same set of values for this set of attributes (the "uniqueness" property) and furthermore no proper subset of that set of attributes will suffice to ensure tuple uniqueness (the "irreducibility" property)
4. must satisfy any relevant combination of constraints to which it is subject. Less formally, entered data must obey the logic implied by the business rules (or "constraints") applicable at the point of entry
5. that satisfies the above four rule components will be entered into a relation and will then have all possible consequential entries derived immediately thereafter. Less formally, if it is possible to infer any tuple (or "record") of data values in any other part of the system, based on the system definition and the current data content (this particularly to include the entry just made), then all such inferences should be drawn. This process is known as "consequence closure".

Statements true within the Theory: Axioms

There are four statements whose truth must be accepted as self-evident within this approach to specification. These serve to define a minimal part of the system. The remainder of the system is then constructed by applying the inference rule to offered data statements and proving their correctness by resolution.

Statements true within the Theory: Theorems

There are very many statements that are provably true and hence have the status of theorems within the knowledge repository. All of these are built by applying the inference rule to unproven data statements and proving their correctness by resolution. These theorems are later used to support correctness reasoning about the specification of the application when it is presented.

5 Construction of the Toolset

In operation, the user of the toolset prepares a specification and offers it to the reasoning system (colloquially called the "compiler") for checking. If the results are not as expected, then either the specification or the compiler is at fault. One important question must be answered: how do we know that the compiler is correct (and therefore that the fault lies in the specification)?

5.1 Correctness Approach

We can show that the fully built compiler will accept its own specification. While this is a useful first step, it uses compiler theorems to show the correctness of compiler theorems – this element of circularity must be removed.

We believe the compiler to be correct because we have specified it, in its own notation, and shown that each statement in its specification is correct according to the rule of inference and that minimal and *independent* dataset needed to support this reasoning. This independent dataset represents a meta-compiler: a much smaller and more abstract system whose sole purpose is to support the formal and provable construction of the full compiler. This approach allows us to move the correctness problem from the compiler to the meta-compiler, where it becomes more tractable.

We believe the meta-compiler to be correct for similar reasons. We have specified it, using its own notation, and shown that each statement in its specification is correct. However, we here use a separate dataset, containing *only* the axioms of the approach. Each statement is therefore correct according to the rule of inference and the axioms of the theory: the axioms are to be accepted as self-evidently correct. There is thus a layered reasoning approach, so that correctness ultimately depends on the validity of the rule of inference and the four axioms. This information is sufficiently small to be examined in considerable detail if it is necessary to do so.

5.2 Construction of the System

The diagram (Figure 2) shows how the system is constructed. Initially there are a small number of axioms (currently four) that are simply placed in the knowledge repository. These (and only these) support the next level of reasoning. The inference rule is then applied to the meta-compiler data and each offered tuple is proved by resolution before inclusion in the repository. At this point, the repository includes a "knowledge" representation of the meta-compiler; the axioms could be removed because every proof that might require to reference them will instead reference a lower-level theorem, preserved as a lemma[3] at the meta-compiler level, in the knowledge repository.

The inference rule is now applied to the specification data for the compiler and, as before, the correctness of each offered tuple is proved by resolution (with reference to all the meta-compiler theorems, used as lemmas) prior to being included in the repository. Again, the repository now includes a representation of the compiler; the meta-compiler could be removed because every proof that might require to refer to this knowledge will instead reference a lower-level theorem, preserved as a compiler lemma, in the knowledge repository.

The inference rule can now be applied to the specification data for the application being specified and, as before, the correctness of each offered tuple is proved by resolution (now with reference to all the compiler theorems, used as lemmas) prior to being included in the repository. As before, the repository now includes a representation of the application; the compiler could be removed because every proof that might require to reference this knowledge will instead reference a lower-level theorem, preserved as an application lemma, in the repository.

This approach means that every step in the build process is provable and traceable, back to the original axioms through the sequence of intermediate proof steps and lemmas. It also means that every step in the reasoning processes for the application specification is similarly provable and traceable.

---

[3] lemma: a preliminary proposition; or a premise taken for granted. Used here to indicate a theorem already proved which is used as a component in a further proof.

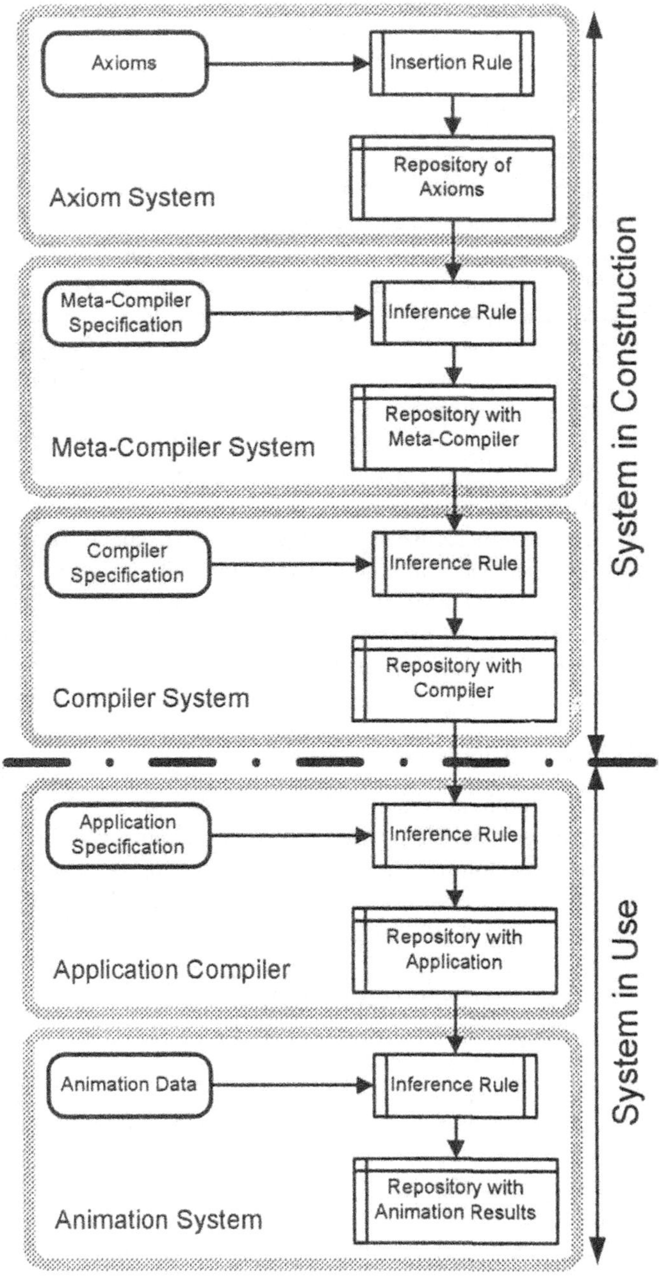

Axioms
Insertion Rule
Repository of Axioms
Axiom System
Meta-Compiler Specification
Inference Rule
Repository with Meta-Compiler
Meta-Compiler System
Compiler Specification
Inference Rule
Repository with Compiler
Compiler System
Application Specification
Inference Rule
Repository with Application
Application Compiler
Animation Data
Inference Rule
Repository with Animation Results
Animation System
System in Construction
System in Use

Figure 2

In each step in this multi-level process, the inference rule essentially activates what has previously been built. Therefore, if the "last" build was the application specification, we can offer data (since that is what the application expects), and activate the application with that data, using the inference rule. This will illustrate the functional behaviour actually captured in the specification, which may or may not be the same as either what the specifier intended or what the client wanted! There is an analogy to testing the correctness of a program with data, but here it is the correctness of the behaviour defined by the specification that is being tested by reasoning: that is, without any code having been written.

If the application compilation completes without error, then the inference rule can be applied to the animation data for the application. As before, the correctness of each offered data tuple can be proved by resolution prior to being included in the repository. This is the final "animation" step. Each proof may refer to any of the application theorems, used as lemmas. At this stage, of course, we do not attempt to remove the representation of the application. Because the same process is used, only the data being different, even the animation is provable and traceable back to the rule of inference and the original axioms through the sequence of intermediate proof steps and lemmas.

Our experience suggests that specifications under-pinned by formal methods are no more accurate than programs when first written. It should therefore be expected that a goodly number of errors will be exposed by the compilation and animation processes. These errors would have to be located and corrected in any event, so no extra work is involved; but correcting these errors early, and before any code is written, is much less expensive than correcting them later.

6 Specification Development

A model is a simplified version, obtained by abstraction, of the corresponding real world system. A model is usually built as a preliminary representation of the system, and is to be followed during the construction of that system. Models operate according to certain scientific principles, and these may be abstract. Specification is concerned with building specialised models of real world systems. Given a theory of specification and a reasoning system constructed according to the theory, we can now construct better specifications.

6.1 User Interface

If mathematical formality is to be concealed from the user, and the formal language also, then the user dialogue must obtain the information needed by the translator from the user, so that the formal text can be produced internally. The user interface operates a "form-filling" dialogue to collect the required values for particular parts of the formal text.

For example, a new attribute within the system is defined by its name and a number of properties, some of which are mandatory while others (for example, the *Attribute Description*) are optional. One part of an example dialogue is encapsulated in the following example:

Attribute Name	patient_NHS_number
Attribute Class	patient
Attribute Key Type	primary
Attribute Index Type	unique
Attribute Data Type	nhs_number
Attribute Description	The unique number by which a patient is known.

The user completes the right-hand column only. Wherever possible, assistance is provided by the provision of pull-down lists from which a value can be selected, rather than forcing the user to enter a pre-existing term. For example, *Key Type* is system limited to an enumerated set of values from which one can be selected; *Data Type* is system and application dependent (jointly) and current entries can be displayed for selection; and *Attribute Class* is entirely application dependent and may or may not have been previously defined. Where application dependence exists, and the prerequisite entry is not (yet) present, a tentative entry is made for the name only; this can later be completed by the user.

Any form can be completed either in its entirety or else in part only. In this latter case, later inclusion of any omitted mandatory values will be required. The omission of mandatory values anywhere in the dialogue will cause errors when compilation is attempted. A blank form is presented for each new attribute; partially completed forms can be recovered so that additional information can be added.

Comparable forms exist for all other concepts; for example, the definition of user-defined data types (such as *nhs_number* in the above example) is permitted and is undertaken through a similar dialogue. A library of functions is provided; the user can either use these directly or else use them as components of more complex user-defined functions.

The use of a "form filling" approach may not be the most "user friendly" approach but it is simple to learn and use. It provides a pattern with which users can readily become familiar. Within the toolset, we take advantage of this pattern to avoid some of the language processing problems of a free format language and there is no ambiguity in the translation to a formal notation. Ease of translation and ease of use are substantial advantages and our over-riding objective was to demonstrate that formal methods can be concealed while still being used reliably. This form of dialogue could easily be changed on a modular basis if desired.

6.2 Specification Correctness

The CREATIV product provides a workbench that accepts specification data as input. The workbench minimally constrains the way in which data is entered; this can be done in (almost) any way that the user may wish. One of the translator tasks is to reorder the data internally into a logical sequence prior to compilation, so that "simpler" theorems are proved before they are needed as lemmas in the proof of more "complex" theorems.

The specification overview is presented as a diagram; the analyst is supported by more detailed tabulations of additional information. The diagram is similar to, but not the same as, the Business Activity diagram of UML. An example diagram is described later (Section 6.3).

The developing version of the specification can be submitted to the compiler at any time. The specification need not be complete. It is advisable to do this frequently, so that errors can be detected, diagnosed, and corrected as early as possible and before subsequent work has built on possibly erroneous foundations. CREATIV provides a diagnostic for each error encountered, presented in terms of the specification currently under construction. Only specifications that are error-free can be animated.

If the current version of the specification is error-free, then it can be submitted to the animator. As before, the specification need not be complete. Again, it is advisable to do this frequently because errors in the specification are easy to make. It is then possible to test and correct the functional behaviour captured by the specification, so that the growing specification remains entirely correct and tested except for the area currently under enhancement. This approach not only allows the client the opportunity to observe and comment on the correctness of the specification from his viewpoint, which is different from that of the analyst, but also makes the progress of the specification activity tangible and measurable.

Using this approach, the preparation of a specification iterates repeatedly through the following activities:

- *data entry* into the workbench, until the specification is *complete*
- *compilation* by the reasoning system, until the specification is *consistent*
- *animation* by the reasoning system, until the specification is *correct*

The specification will be *comprehensible* throughout because no access is given to the internal and formal notation. The only information that is provided at the specification input stage is the overview diagram, which for complex systems may be quite large; this is supported by more detailed tabulations that record the information entered.

It is likely that the analyst will acquire the information on which the specification is to be based and will probably carry out not only the data entry but also the compilation, correcting any errors and inconsistencies. Queries are likely to be resolved with the relevant client representatives. It is usual for these activities not to be obvious to the client. However, once one or more animations are known to run to completion without apparent error, it is of prime importance to confirm these "scenarios" with the client. This ensures good client involvement with the ongoing specification activity, confirms that the scenario behaviour is correct, and may suggest new scenarios to be tested. Quite separately, it also allows the progress of the specification activity to be measured.

6.3 An Example Specification

As a small example of the overview diagram, we offer the following specification of the operation of a lending library. In its initial form, this is shown on the centre (enclosed) part of the following diagram (Figure 3). The library purchases items (these might be books, records, or cassettes) that then become available for lending as "available stock". The library also maintains a list of members, with relevant details for each member; this list represents each "current member". The library will accept a "borrowing request", but requires that this must be made by a member (R09) and that this must be for an item of stock that is available (R07). If these two constraints, or business rules, are not satisfied, then the borrowing request is refused and the system returns to its original state.

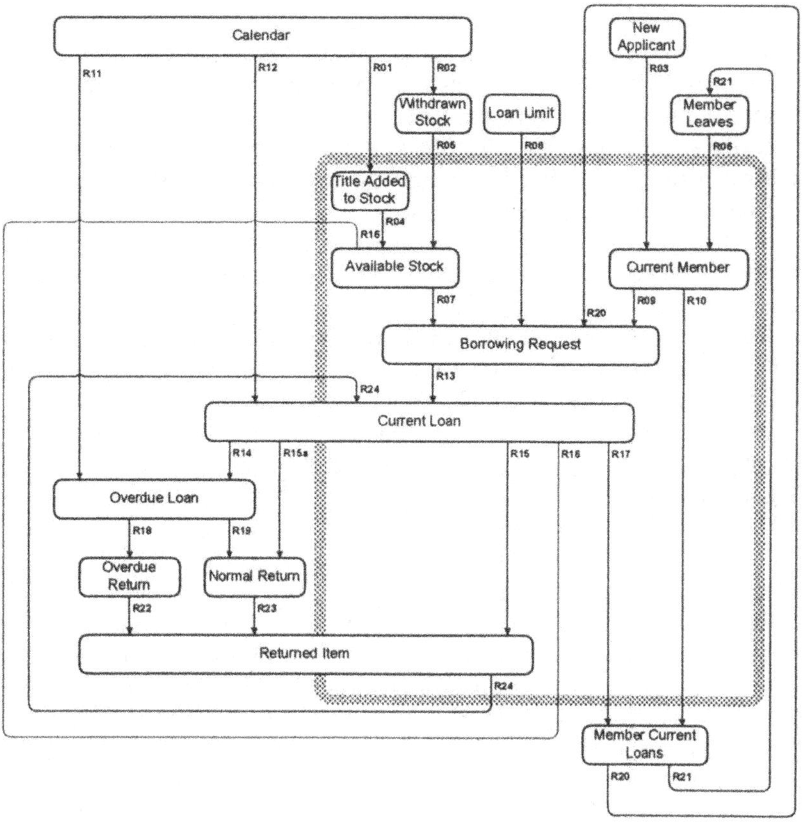

Figure 3

If the request does satisfy the constraints, then the borrowing request is accepted and, as a consequence, the loan is approved and made (represented by an entry in "current loan"). One necessary consequence of making the loan is that the item becomes unavailable to other borrowers, pending its return (a physical constraint effected by R16). Upon return, a "returned item" must match the corresponding "current loan" item; the return of the item not only cancels the member loan (effected by R24) but also restores the availability of the item to other borrowers (through R16).

This very simple model is inadequate as a model of a lending library. But, simply because it is easy for users to understand, it is comparably easy to ask informed questions about its deficiencies:

- how do we limit the number of loans to one borrower?
- how do we represent members who leave the library?
- how do we levy fines on overdue items when returned?
- how do we show the withdrawal of elderly stock items?

All these facilities, and others, can be included and some are partly shown in the outer part of the model. When models (specifications) are extended, some other (possibly unforeseen) issues may arise as a result. For example "current member" now becomes the set of all applicants who have joined less all those who have left. The analyst may discuss with the user whether a member may join and leave several times (since this may affect the key structure), or under what conditions a member is permitted to leave. It is clearly not desirable that he should leave before he has returned all his borrowings and so R21 (perhaps not previously anticipated) enforces that "member current loans" must be zero before the member leaves. Similarly, the notion of "available stock" now becomes the set of all those items that have been added less all those that have been withdrawn. The issue of whether some items may be replaced after withdrawal needs to be considered (since this again may affect the key structure). Also, there is currently no check in this model of the physical necessity for withdrawn items to be available (not on loan) when they are withdrawn; an additional constraint on "withdrawn stock" confirming the physical presence in "available stock" of the item to be withdrawn would be necessary in reality.

If it is not already apparent, it should be noted that the lines (constraints or business rules) do not all behave in the same way. These differences are necessary and form part of the specification detail.

With a little practice, it is not hard to construct such models of the information systems needed by a business, whether these are for systems that are already present or those that will be required in the future. Although generally similar to the "business activity" models of UML, they will contain substantially more detail when they are complete. The advantage of such models is that users and reviewers have no difficulty in comprehending what the model represents in the real world. The possible lack of detail at the start of the specification development is invariably supplemented by the reader's knowledge and common sense, and this is likely to be made available to the analyst during informal review and discussion. When the detail is added later, the actual consistency and behavioural correctness of the specification can be demonstrated.

6.4 Specification Comprehension

Different people have different viewpoints on the same real world system, and it is therefore useful to be able to make different presentations of the same specification to different people. This is achieved by presentation modules (translators) that generate specific presentations from the one underlying repository. A number of translators have been constructed, and are embodied within the toolset; some others have been experimental.

Overview Diagram

The diagram, one example of which has already been shown in Figure 3, is the principal presentation for many users. It relates well to users' understanding of their business, it is informative and easy to comprehend, and it aligns well with animation results. For many people, this is sufficient.

The diagram is produced by the workbench, with some icon placement information being provided by the analyst. We consider it important to exercise a measure of control over the layout, so that the diagram retains business coherence for users. If the diagram is too large to fit on one sheet of paper, a "tiled" output is produced that can be scaled to a size determined by the user.

Data Tabulated by Input or Class

For those who require more detailed technical information, this can be extracted from the repository and tabulated. These tabulations can be provided either to align with the workbench dialogues (these are designed to support the data entry activity and to ease subsequent error correction, if necessary) or to align with the class structure of the application (these are designed to support technical review).

Experimental Translators

We have investigated the production of a translation from the repository to the Z notation as well as the production of specification information in other notations (entity relationship diagrams, data flow diagrams, and SSADM entity life history diagrams). These can all be produced, though they are not included within the present version of the CREATIV toolset.

6.5 Interfaces to Collaborative Systems

If program code is to be written by hand, then we would wish that such code be written once only and be as nearly correct as possible. This is because the coding process is labour-intensive and therefore expensive; for this reason, rewriting should be avoided (this is one disadvantage of prototyping approaches to system building). The coding process is also error-prone; for this reason, as little corrective rework as possible should be necessary. The provision of a good specification contributes materially to the achievement of each of these two objectives.

However, if program code can be machine-generated, then expensive rewriting and error-prone code correction can both be avoided. This allows a prototyping approach to be more justifiably applied, but now at the specification level. If the specification is written in a language sufficiently formal that code generation can be supported, then the initial specification can be less good and used for code generation; errors identified by testing the running code can then be corrected in the specification. This is iterative "rapid prototyping" at the specification level rather than the code level.

We adopted the first approach; another company adopted the second. In both cases, it is important to note that the specifications (our output and their input) are written in formal languages; automatic translation between the two notations is therefore possible. If we determine user requirements and formalise a specification in our notation, and they accept a (formal) specification in their notation, then we can collaboratively proceed from user requirements to executable code by rule and with very little manual involvement (and therefore very few errors), provided only that a translator exists between the two formal languages.

The CREATIV product provides a translator that allows direct interfacing with the Perfect Developer toolset from Escher Technologies Ltd (Crocker, 2004). The Perfect Developer toolset accepts specifications written in their own notation (the Perfect language) and allows code (Java, C++, or Ada) to be generated directly. This means that once the client requirement has been captured and is agreed to be correct, we can translate the requirement to a specification in the Perfect language and then progress to a fully running system, by rule and with almost no manual intervention.

The CREATIV approach uses reasoning to test the specification itself by animation. In contrast, the Perfect Developer approach uses rule-based code generation to support a rapid prototyping approach: specify, generate, and test, correcting any errors only by revising the specification. We suggest that the two techniques are complementary, rather than mutually exclusive: animation should be used first, since this can be done very early in the project, during the elicitation of requirements and before screens and databases are designed. This may also show some of the perceived requirements to be mutually incompatible. Later, prototyping by code generation should be used as well, since the generated code will eventually form the basis for the delivered product. The use of code generation will also provide an early indication of response time and performance.

The Perfect Developer toolset also contains a theorem prover (not resolution based) that allows users to enunciate *properties* that they intend the proposed system to possess and then determines whether those properties are indeed present. It is already evident that we attach considerable importance to specification correctness; we therefore deduce as many proof obligations as possible from the specification data stored in the repository so that we can obtain independent proof of their correctness using the Perfect Developer theorem prover. We expect all our generated obligations to be successfully provable. Also, we support the inclusion of an unlimited number of additional proof obligations, provided by the user, to extend the proof coverage given to the specification.

6.6 Practical Experience with CREATIV

In this section, we have so far described how the CREATIV toolset is used; we now describe some practical experiences of its use.

Database Design Problems

One project, undertaken for a Government department, required the analysis of a database system built by an independent supplier. The system was defined in an SSADM specification of high quality that had been independently reviewed, but the specification was not mathematically formal. The database as built contained many normalisation errors that would have led to many operational problems had the database entered service. We used CREATIV to specify this aspect of the system and were able to provide a convincing demonstration (and a mathematical proof) of its unsuitability for service.

We have subsequently undertaken a similar project for a separate commercial organisation, though here the problem was made worse by the absence of a clear specification against which the database might have been designed.

Quality Assurance

One "proof of concept" project, undertaken for a government department, required an existing specification (again of high quality but not mathematically formal) to be subjected to rule-based analysis in order to confirm its quality. The reason for this additional quality examination was that the system was intended to have a high level of public access. The project was valuable to the client since four errors were discovered quickly. These included an error in a stated business rule, an error in a pair of rules that were mutually incompatible, an error where the ordering of rules could lead to invalid data submission, and one data type incompatibility. The data type error would probably have been exposed during implementation; whether the three rule errors would have been found is less clear. Given that two of the errors relate to a pair or a sequence of rules, very careful and thorough testing would be needed to guarantee their exposure.

Database Corruption

CREATIV has been used to analyse a small database system that was to be replaced. The replacement was justified for a number of reasons, but one factor was the infrequent and erratic corruption of parts of the database. Analysis of the system showed that this arose from an interaction between different parts of the database that would only occur with a particular combination of actions by two different users. A CREATIV specification for this part of the system explained the problem very clearly; furthermore, the proof obligation for the user's expectation of what should happen could not be satisfied (was unprovable) for precisely the cause that had been identified.

Parallel Specification

The Software Requirements Definition (SRD) for the replacement system (above) was written by an independent supplier, while we undertook the preparation of a parallel CREATIV specification. A portion of this had been successfully tested by animation. However, the above proof obligation remained unprovable, because of the existence of a further and more subtle problem. This problem, related to the sequencing of entries to the database, would also have led to unexpected behaviour though not to database corruption. The CREATIV specification was corrected, the proof was successful, and the change was passed to the supplier to incorporate in the SRD.

At the time of submission of this paper, this project is still ongoing. The CREATIV specification currently spans about 20% of the project; currently, the full overview diagram contains about three hundred icons. We translate this to the Perfect notation on a regular basis; currently, this generates about 14,000 lines of Perfect specification. This includes about 2,700 theorems or properties that the final system should possess; these are generated by rule during the translation process. About 98.5% of these theorems are currently provable by the Perfect Developer toolset. Of the unprovable theorems, about half arise from one known cause (to be corrected in the translator) and the remainder are under investigation.

7 Conclusions

Too many software development projects end in failure. One reason, though not the only one, is the absence of a good specification: one that is complete, consistent, correct, and comprehensible. Formal methods are essential to achieve this, but their presence should be concealed within the toolset so that users do not need formal methods expertise to achieve the advantages of their use.

The use of formal methods mandates a rigorous and scientific approach to specification, so that reasoning about the system can be supported. One possible way, though not the only one, is to formalise specification knowledge as an axiomatic system. If we not only construct the reasoning system but also build and animate the specification in this (axiomatic) way, then all operations in the system are provable and traceable. We have constructed the whole of the reasoning system (though not the rest of the workbench) in this way and demonstrated its consistency and correctness.

More recently, we have successfully used the approach on several government projects. Our experience is that a range of information system projects can be specified using this approach; that specification testing with the client is valuable; and that the use of proof reveals further issues not exposed by animation testing. It is important to note that proof has a "coverage" characteristic (like test coverage): it is perfectly possible for a system to be subject to successful proof while still containing errors if the proof obligations did not encompass each error.

In conjunction with the Perfect Developer toolset from Escher Technologies Ltd. it is possible to generate executable code directly from the specification (with optional manual refinement if required). It is also possible to generate a very substantial number of proof obligations by rule and to discharge these entirely automatically except in those cases where the obligation is genuinely in error; this provides independent proof of the properties of the generated system.

8 References

BCS Review (2001):
http://www.bcs.org.uk/review/2001/html/p061.htm

Chaos Report (1994):
http://www.standishgroup.com/sample_research/chaos_1994_1.php

Crocker (2004):
"Safe Object-Oriented Software:
The Verified Design-by-Contract Paradigm", D. Crocker,
Proceedings of 12th Safety Critical Systems Symposium,
Springer-Verlag (London)

Pierce (2002):
"Types and Programming Languages", Benjamin C. Pierce,
MIT Press (2002), ISBN 0-262-16209-1
Page xxi, quotation attributed to Tom Melham

MANAGING RISK IN THE RAILWAY INDUSTRY

Development and Use of the UK Railway Network's Safety Risk Model

Author
Colin Dennis, BSc(Hons) CEng MIMechE
Head of Risk Assessment,
Rail Safety and Standards Board,
Evergreen House,
160 Euston Road,
London.
NW1 2DX
e-mail: colin.dennis@rssb.co.uk

Abstract

The Rail Safety and Standards Board (RSSB) has a responsibility to lead and develop long term safety strategy and policy for the UK mainline railway network. One of the key objectives of the strategy is to reduce risk on the railway network by controlling the hazardous events and the precursors to the hazardous events that can occur. It is essential, however, that the control of the risk is carried out in an open and explicit manner, so that end users in the industry can be assured the controls imposed are effective and do not cost disproportionately more than the benefits they provide.

The Safety Risk Model (SRM), which is a detailed fault tree and event tree analysis model, has been developed by RSSB to provide a structured representation of the cause and consequences of potential accidents arising from the operation and maintenance of the railway.

The paper describes how the SRM has been developed up to the issue of version 3 in February 2003 and the plans for future development.

1 Introduction

RSSB has a responsibility to lead and develop long term safety strategy and policy for the UK mainline railway network.

This responsibility is met through the production of Railway Group Standards (RGS), RSSB approved codes of practice (RACOP), Guidance Notes, production of the Railway Group Safety Plan (RGSP) and through its current risk assessments and the Safety Management Information System (SMIS). The purpose of these activities is to reduce risk on Network Rail Controlled Infrastructure (NRCI) by controlling the hazardous events. It is essential, however, that the control of the risk and hazardous events on NRCI is carried out in an open and explicit manner, so that end users in the Railway Group can be assured the controls imposed are effective and do not cost disproportionately more than the benefits they provide.

The Safety Risk Model (SRM) has been developed by RSSB to provide a structured representation of the causes and consequences of potential accidents arising from railway operations and maintenance on the NRCI and in other areas where RSSB has a RGSP commitment to record and report accidents. RSSB's objectives, which drive the SRM, are summarised in Section 2.

The SRM consists of a series of fault tree and event tree models representing 122 hazardous events that, collectively, define the overall level of risk on NRCI.

The SRM relates to the systemwide risk on NRCI covering all running lines, rolling stock types, locations and stations currently in use. Version 3 of the SRM, does not cover the non-NRCI related risk associated with yards, sidings, depots and station car parks. Risk profiles for specific lines of route and train operating companies, are not available currently within the SRM.

The SRM has been designed to take full account of both the high frequency low consequence type events (events occurring routinely for which there is significant quantity of recorded data) and the low frequency high consequence events (events occurring rarely for which there is little recorded data). The results for each hazardous event are presented in terms of the frequency of occurrence (number of events/year) and the risk (number of equivalent fatalities per year).

It should be noted that the results from the SRM represent the residual risk predicted currently on NRCI i.e. they represent the level of risk assuming all current control measures are in place with their current degree of effectiveness. No overall risk prediction excluding the influence of all current controls is possible.

2 Safety Risk Model Objectives

The primary objectives of the SRM are to provide:

- An understanding of the nature of the current risk on the NRCI.
- Risk information and risk profiles relating to the NRCI, the Railway Group and the wider railway industry.

This will:

- Assist in the development and validation of Railway Safety Cases within the Railway Group.
- Provide risk information for use in risk assessments throughout the Railway Group.
- Assist in developing priorities for the Railway Group Safety Plan.
- Identify and prioritise the revision of Railway Group Standards, in terms of their contribution to risk mitigation.
- Enable ALARP assessments and cost benefit analyses to be carried out (i) to assist in the decision making process regarding the merits of technical changes or modifications, and new infrastructure investment and (ii) to assist in the development of safety justifications for proposed changes to RGS.
- Assist in identifying additional control measures, which would reduce risk.
- Aid the understanding of the contribution of a particular item of equipment or failure mode to the overall risk.
- Assist in the identification and prioritisation of issues for audit.
- Prioritise areas for safety research.

3 Background

The SRM is based on the quantification of the risk resulting from hazardous events occurring on the NRCI that have the potential to lead to fatalities, major injuries or minor injuries to passengers, staff or members of the public. In the context of the SRM a 'hazardous event' is taken to mean an event that has the potential to lead directly to death or injury.

For each hazardous event there could be a single or combination of precursors (system failures, sub-system failures, component failures, human errors or physical effects) that could result in the occurrence of the hazardous event. For example, a derailment would be considered to be a hazardous event as it can lead directly to injuries, whereas a broken rail would be classified as a cause precursor because without the occurrence of a subsequent derailment, no injury would occur.

The SRM hazardous events adopt the incident categories defined by Her Majesty's Railway Inspectorate (HMRI) *ie,* hazardous events, train accidents (HET), movement accidents (HEM) and non-movement accidents (HEN) relevant to the operation and maintenance of the NRCI, where:

- train accidents are accidents to trains and rolling stock.
- movement accidents are accidents to people caused by the movement of trains but excluding those involved in train accidents.
- non-movement accidents are accidents to people on railway premises but not connected with the movement of railway vehicles.

Over the years there have been several comprehensive hazard identification studies carried out on railway systems in the UK that could be utilised to avoid the need for further detailed hazard identification assessments. It has, however, still been necessary to develop a full list of hazardous events and their associated precursors for inclusion in the SRM.

Traditionally, hazard identification has considered whether fatalities or injuries to people occur as a result of equipment, system or procedural failures. However, to ensure that all possible events leading to fatalities or injuries were identified for inclusion in the SRM, in addition to referencing the existing hazard identification studies it was decided to consider the hazards using a novel approach whereby the generic injury mechanisms by which a person can be killed or injured are identified. The question was then asked 'how can this type of injury be caused on a railway system?'

In order to give confidence in the completeness of the hazardous events identification process, the derived list of hazardous events was cross-checked against the previous hazard identification studies undertaken on the UK railways.

The high-level list of hazardous events derived from the above process was sub-divided into events where the frequency or the consequences of each event are significantly different. This process resulted in the definition of 122 hazardous events that form the basis of version 3 of the SRM. There is a separate cause/consequence model for each hazardous event. Examples from the list of hazardous events are presented in Table 1 at the end of the text.

To provide a greater level of detail in the modelling process, in some cases it has been necessary for individual hazardous event models to include further subdivisions where the precursors, frequency of failure or the overall consequences could be significantly different. Examples of such further subdivisions are as follows:

- Hazardous event HET-01 *Collision between two passenger trains (other than at platform)* has been broken down into the speed of collision (fast and slow) and the type of collision (rear end, head-on and side-on) and

- Hazardous event HET-12 *Derailment of a passenger train* has been broken down into the speed of derailment (fast and slow) and the location of derailment (open track, single track tunnel, twin track tunnel, bridges/viaducts and stations).
- Existing hazard identification studies were also used to identify and extract the precursors and the control measures and mitigation factors which are applicable to each hazardous event. This information forms the basis of the development of the SRM for each hazardous event.

4 Modelling Approach

The SRM has been developed in the form of a cause and consequence analysis using fault trees and event trees to represent each of the hazardous events.

"Risk" in the context of the SRM relates purely to safety risk in terms of an estimate of the potential for harm to passengers, staff and members of the public from the operation and maintenance of the railway. The risk (known as collective risk) is defined as the average number of fatalities or equivalent fatalities per year that would occur if the system was operated for an extended period of time. The average risk over an extended period of time must be considered in order to capture the low frequency high consequence events in addition to the high frequency low consequence events. In the case of train accidents, for which there are potentially a significant number of low frequency high consequence events, this could relate to many hundreds of years of operation. For movement and non-movement hazardous events, which tend to be dominated by high frequency low consequence events, this would normally relate to tens of years of operation.

The risk associated with a particular hazardous event is calculated in terms of:

Frequency

The average frequency at which the hazardous event occurs

(*eg* number of events/year)

X

Consequences

The average consequences should the hazardous event occur

(*eg* the number of fatalities or equivalent fatalities/event)

=

Collective Risk

(*eg* the average number of fatalities or equivalent fatalities/year)

Equivalent fatalities are a measure used to make allowance for the potential impact of major injuries and minor injuries when carrying out cost benefit analyses. Ten major injuries and 200 minor injuries are both equal to one equivalent fatality.

The overall risk for the railway is the sum of the risk for each individual hazardous event.

The frequency of each hazardous event is modelled using fault tree analysis. The possible outcomes from each hazardous event are modelled using event tree analysis.

During the development of the SRM it was recognised that NRCI has many differing characteristics relating, for example, to the different train operating companies, the variety of trains and different service intensities, which will affect the estimates of risk in terms of differing causes, frequency of failures and consequences of hazardous events. This level of detailed modelling is beyond the scope of the current SRM. The SRM is therefore a systemwide model to support RSSB in determining systemwide risk and risk controls. Nevertheless, in relation to train accidents the systemwide SRM includes:

- the passenger loading (night, off-peak, peak and crush loaded)
- the train speed
- whether secondary effects occur such as collision with a train on an adjacent line, fires, collisions with lineside structures or structural collapse onto train.

A diagrammatic representation of the relationship between the fault tree analysis and the event tree analysis in the context of the overall SRM is presented in Figure 1 below and Figure 2 at the end of the text.

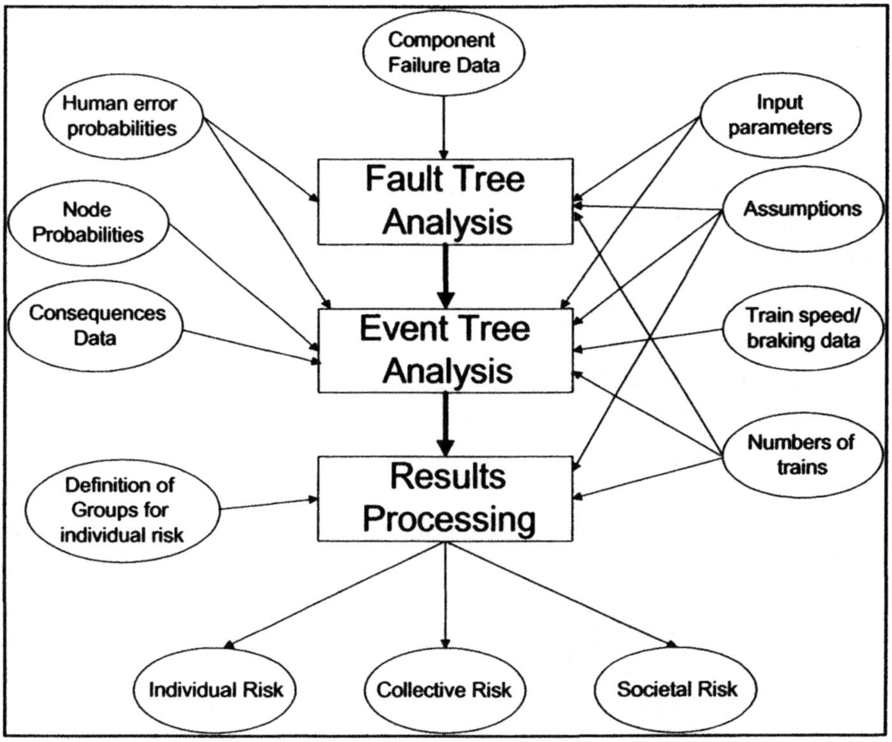

Figure 1: Information flows within the SRM

Figure 2 shows the basic calculations that underpin each of the 122 hazardous event models within the SRM.

The SRM enables risk to be calculated in terms of:

- Collective risk - the average number of equivalent fatalities per year occurring on the NRCI
- Individual risk - the annual probability of fatality per year for a particular passenger or staff group using the railway
- Societal risk - a measure of the frequency of accidents that lead to multiple fatalities.

4.1 Input data

The fault trees and event trees have been populated with input data relating to the frequency or probability of component, equipment, system, and human failures, the probability of circumstantial events occurring and the consequences of each fault sequence.

Data was obtained from a wider variety of industry data sources including:

- SMIS (Safety Management Information System)
- Signal Passed at Danger (SPAD) analysis (derived from SMIS)
- Train fires database
- RAILDATA (broken rails database)
- Derailment database
- Signalling & Telecommunications equipment failure database
- Her Majesty's Railway Inspectorate database/annual reports
- Generic failure and reliability data sources.

Where data was not available for the quantification of the cause and consequence precursors, use was made of:

- Human error probability assessments using the Human Error Assessment and Reduction Technique (HEART)
- Expert judgement from in-house expertise within RSSB and Network Rail.

For cause precursors where no data was available, where possible the following techniques have been used to assign frequencies or probabilities:

1. Expert judgement from technical specialists within RSSB
2. A statistical method using the χ^2 (chi) distribution for zero failures (Green and Bourne 1977).

There is clearly the potential for significant uncertainty in input data used in the analysis, particularly where the data are based on expert judgement. However, in developing and reviewing the SRM steps are taken to try and minimise the amount of uncertainty as discussed in Section 8 below.

5 Model uses

The SRM represents a comprehensive systemwide computer based model of the UK railway network that can be used to meet the objectives defined in Section 2. The model has been in use since January 2000 and enables risk to be examined in a number of ways:

a) Graphical representation of the risk profile for all groups or individual hazardous events.
b) Undertake "what if" analysis in relation to assessing the effects of changes in the precursor data, the introduction of new or modified risk control measures, degradation in existing risk control measures, increases and decreases in average passenger loadings, and the removal and introduction of rolling stock types, etc.
c) Determine the risk contribution and risk profile, from individual, or groups of precursors.
d) Determine the frequency and average consequences per event for each of the hazardous events.
e) Determine the relationship between the frequency of occurrence and the number of fatalities for all groups of or individual hazardous events.
f) Enable cost benefit analysis assessments to be undertaken.
g) Provide the basis for assessing the risk for a particular railway route or for a particular train operating company.
h) It should be noted that in order to assess the risk for a particular line of route or for a particular train operating company ((g) above), each model, each assumption and data input to the SRM, must be examined for the relevance to the particular railway route or for a particular train operating company.

6 Model development

The results from version 1 of the SRM were presented to the Railway Group in a document called the Risk Profile Bulletin (Dennis C et al 2001a) in January 2001 with version 2 following shortly afterwards in July 2001 (Dennis C et al 2001b). Version 2 incorporated some minor modelling changes but the main emphasis was on improving the presentation of the results within the Risk Profile Bulletin.

RSSB has a commitment to update the SRM annually to take account of the latest data relating to the operation and maintenance of the railway. As the SRM and the results from the SRM were used it became apparent that there were a number of items that would require a more significant revision of the model in addition to the data update. These drivers for change were as follows:

1. The introduction of the Train Protection and Warning System (TPWS) to help prevent trains colliding if a red signal is passed at danger.

2. Advancements in the modelling of the consequences of train collisions using detailed finite element analysis models.

3. A recognition that in some areas, most notably non-passenger train derailments, it was known that level of modelling used in versions 1 and 2 of the SRM had led to pessimistic estimates of the overall risk on the railway.

4. The need to expand the scope of the model to include the risk contributions from on-train incidents and all incidents occurring on stations that had not been covered in version 1 and 2.

5. The need to include the risk contribution from all minor injuries. Versions 1 and 2 of the SRM only included the RIDDOR (Reporting of Injuries, Diseases, and Dangerous Occurrences Regulations) reportable minor injuries (Health and Safety Executive 1995).

6. The need to establish the risk contribution from the removal of the older style Mark 1 slam door rolling stock by the end of 2004.

The most significant changes relating to version 3 of the SRM are therefore:

- Update of the data used as the basis of the analysis to the most recent data available.
- The inclusion of the risk reduction implied by the incorporation of the TPWS that is to be installed across the network by the end of 2003.
- Improved collision consequences modelling, for the collision-related hazardous events.
- More refined modelling of non-passenger train derailments to include separate modelling of freight trains on passenger lines, freight trains on freight only lines and ECS (Empty Coach Stock) and Parcels train derailments on passenger lines.
- The inclusion of all minor and shock/trauma related injuries not just the RIDDOR reportable minor injuries (Health and Safety Executive 1995).
- The inclusion of the risk contributions from on-train incidents and all incidents occurring at stations.

7 Results

The total risk to passengers, staff and members of the public from the 122 hazardous events modelled within the SRM with TPWS but excluding suicides, is predicted to be 203 equivalent fatalities per year. A breakdown of the total risk by accident category is shown in Table 2:

Accident category	Risk (eq. fats./yr.)		
	SRM version 3		SRM version 2
	Without TPWS	With TPWS	Without TPWS
Train accidents	17.5	14.9	24.4
Movement accidents (excluding trespass)	63.1	63.1	24.8
Non-movement accidents (excluding trespass)	63.8	63.8	27.5
Trespass	62.0	62.0	61.2
Total	**206.4**	**203.8**	**138.0**

Table 2: Total risk by accident category (excluding suicides)

There are a number of key factors that cause the differences in the predicted levels of risk between version 2 and version 3 of the SRM:

Train accidents:

- Update of the data used as the basis of the analysis that resulting in a reduction in the number of train-to-train collisions, buffer stop collisions and derailments predicted per year.
- Improved collision consequences modelling for the collision related hazardous events that in general lead to a lower number of equivalent fatalities per collision than the levels derived for version 2.
- More refined modelling of non-passenger train derailments to include separate modelling of freight trains on passenger lines, freight trains on freight only lines and ECS and Parcels train derailments on passenger lines. This removed the known pessimism in the version 2 of the SRM leading to a 35% reduction in the level of risk estimated for HET-13 "Derailment of a non-passenger train".

Movement and non-movement accidents:

- The inclusion of all non-reportable minor and shock or trauma related injuries that resulted in a dramatic increase of about 50 equivalent fatalities. This relates to around 10,000 additional minor injuries.

- The inclusion of the risk contributions from on-train incidents and incidents occurring on stations.
- Update of the data used as the basis for the analysis.

In order to identify which hazardous events provide the most significant contributions to the overall risk, a risk profile showing all the hazardous events (including trespass but excluding suicides) that have a risk contribution of greater than 1 equivalent fatality per year is shown in Figure 3 at the end of the text.

It can be seen that HEM-25, *adult trespasser struck/crushed while on the mainline railway*, provides by far the greatest risk contribution at 44.5 equivalent fatalities per year representing 21.6% of the overall risk on the mainline railway.

The highest train accident related hazardous event is HET-10, *passenger train collision with road vehicle on level crossing*, at 3.9 equivalent fatalities representing 1.9% of the overall risk on the mainline railway.

HET-01, *collision between two passenger trains (other than at platform) (without TPWS)*, is the 20th most significant hazardous event with a risk contribution of 2.7 equivalent fatalities representing 1.3% of the overall risk on the mainline railway.

With the exception of the trespasser related events, the main contributors to risk are dominated by the high frequency low consequence events that tend to lead only to non-reportable minor injuries.

When reviewing the results from the SRM it must be recognised that the overall risk contributions for the hazardous events are made up from different profiles of frequency and consequences. The movement and non-movement accidents tend to consist of high frequency low consequence type events *eg* slips, trips and falls, while the train accidents tend to have a significant risk contribution from the low frequency high consequence type events *eg* passenger train derailment where coaches fall on their side and there is a secondary collision with another passenger train on an adjacent line. The effect of these low frequency high consequence events is to increase the risk contribution for the hazardous events above the level that may have been seen in practice over say the last ten years.

One of the key developments of the version 3 model was the detailed modelling associated with the fitment of TPWS to the railway network. The SRM was developed to enable the effects of TPWS to be switched on and off to allow the risk reduction from TPWS to be demonstrated clearly. Table 3 at the end of the text shows the results from the SRM for the without and with TPWS cases.

One of the objectives in the Railway Group Safety Plan (RGSP) (Railway Safety 2003) requires the industry to reduce the benchmark risk arising from Signals Passed at Danger (SPADs) on or affecting the running line in 2000/01 by 80% by March 2009. Analysis of the risk relating only to SPADs from the SRM (*ie* excluding derailments due to overspeeding and buffer stop collisions) suggests that full TPWS fitment will eliminate around 61% of the overall risk arising from SPADs.

A further development that will reduce the risk from collisions and derailments due to SPADs is the removal of older Mark 1 slam door rolling stock that has poorer crashworthiness compared to more modern stock. The SRM has been used to show that the removal of the Mark 1 rolling stock combined with the risk reduction achievable from full TPWS fitment is estimated to give a 67% reduction in the overall level of risk attributable to SPADs. Therefore the SRM has been used to show that the combined effect of full fitment of TPWS and the removal of Mark 1 rolling stock will contribute a large proportion of the overall risk reduction required for the RGSP objective, giving confidence that the industry is targeting the correct areas for risk reduction.

As society in general has an aversion to single accidents that result in multiple fatalities it is important to be able to estimate the potential frequency of accidents which could lead to multiple passenger, staff and member of the public fatalities.

By using the event tree structures within the SRM, for the hazardous events with the potential to lead to multiple fatalities, an overall F-N (frequency versus number of fatalities) curve has been produced. As one would expect the curve shows that as the number of potential fatalities increases the frequency of occurrence reduces rapidly. Key points on the curve are shown in Table 4:

Number of fatalities (passengers, staff and MOPs)	Without TPWS Estimated average number of years between events	With TPWS Estimated average number of years between events	% Reduction due to TPWS
≥ 5	2.4	3.2	25.4%
≥ 10	5.6	7.0	20.6%
≥ 25	20.7	26.1	20.9%
≥ 50	220.1	348.8	36.9%
≥ 100	1204.0	1372.5	12.3%
≥ 200	6979.2	7215.6	3.3%

Table 4: Frequency of incidents leading to multiple fatalities

8 Uncertainty in the results

When producing any risk assessments, which include a significant contribution from the low frequency high consequence type events for which there is little or no historical experience, it is necessary to make predictions on the likelihood and consequences of the specific outcomes following the occurrence of a hazardous event. This process inevitably leads to the requirement for judgement-based assumptions. The accuracy of the overall risk results is therefore sensitive to the accuracy of the assumptions made. All assumptions made within the model are justified and recorded within the project database and associated documentation.

For movement and non-movement accidents, which are based largely on a statistically significant database of incidents, no major assumptions have been made which are likely to have a significant effect on the accuracy of the overall results.

For train accidents however, the results are potentially sensitive to a significant number of assumptions and predictions that are used in the analysis. The level of uncertainty is at present controlled by:

- Wherever possible using recorded data
- Careful selection of subject matter experts for the expert judgement
- Sensitivity analyses to show the effect of particular assumptions on the overall results.
- Benchmarking of the results against known data points e.g. the experience from accidents over the last 10 to 30 years
- Continual sense checks of the results from the model e.g. Does it look right? Is it what I expected?

A separate study is currently in progress to examine the methodologies that can be used for assigning data where little or no records of failures exist for the cause precursors. A PhD project is also being set up to examine the propagation of uncertainty through the model.

Following the completion of versions 1 and 2 of the SRM, it was recognised that in a number of areas e.g. non-passenger train derailments and train collisions, the results from the model were pessimistic. This knowledge led to the requirement for more refined modelling in version 3 of the SRM, thereby reducing the level of pessimism in the model.

The SRM has been subject to peer review by the Health & Safety Executive (Turner 2002) and a detailed statistical review by The University of Strathclyde (Bedford 2003).

9 Uses of the SRM to date

The SRM has been used extensively since the issue of version 1 of the Risk Profile Bulletin in January 2001. Typical uses have included:

- To provide a focus for key risk areas that require further investigation.
- For special topic reports on significant risk issues.
- Advise on priorities for Research & Development projects.
- The preparation of risk assessments for Railway Safety Cases.
- Assessing changes to Railway Group Standards.
- Analysis of accidents that have occurred.
- By technical experts within RSSB.
- Development of a risk assessment methodology for on track plant.

10 Future Development of the SRM

The development and use of the SRM is an on-going project. RSSB are committed to maintaining the SRM up to date and where possible expand and refine the modelling. version 4 of the model is being prepared currently and will include:

- Updating to account for recent data.
- More detailed modelling and improved data analysis to remove known areas of excess pessimism within the model.

Other areas for development being considered are:

- The feasibility of the development of specific "railway route" models.
- Improved level of human factors modelling.

- Use of more sophisticated statistical analysis techniques where only limited data is available for some of the required inputs.

11 Conclusions

The SRM provides a generic model of the safety risk on the UK railway network which:

- Increases the industry's knowledge of the risk and the risk profile from the operation and maintenance of UK railway network
- Allows the identification of areas of railway operation that need further risk controls
- Allows sensitivity analyses to be carried out to determine the risk reduction from the introduction of new control measures
- Allows cost benefit analysis of proposed changes

The SRM has been developed and maintained in house so the models are owned and understood by RSSB.

12 References

Bedford, T, Quigley, J, Walls, L. University of Strathclyde (2003) Statistical review of SRM: WP1 report. June 2003

Dennis C et al, (2001a). Railway Safety. Profile of safety risk on Railtrack controlled infrastructure. SP-RSK-3.1.3.11 Issue 1 January 2001

Dennis C et al, (2001b). Railway Safety. Profile of safety risk on Railtrack controlled infrastructure. SP-RSK-3.1.3.11 Issue 2 July 2001

Green, AE and Bourne (1977), Reliability Technology, March 1977

Health and Safety Executive (1995): Reporting of injuries, diseases and dangerous occurrences regulations (RIDDOR). HMSO

Railway Safety (2003). Railway Group Safety Plan. 2003/2004.

Turner, S. (2002). Health & Safety Laboratory. Review of Railway Safety's Safety Risk Model. RAS/02/11. June 2002.

Table 1: List of hazardous events modelled within the SRM

Haz Event Nunber	Hazardous Event Description
	Train accidents
HET 1	Collision between two passenger trains (other than at platform)
HET 2	Collision between a passenger train and non-passenger train
HET 3	Collision between two non-passenger trains
HET 4	Collision of train with object on line (not resulting in derailment)
HET 5	Collision of train with object large enough to cause structural damage to the train (object above buffer height)
HET 6	Collision between two passenger trains in station (permissive working)
HET 7	Collision between a passenger train and a non-passenger train in station
HET 8	Collision between two non-passenger trains in station
HET 9	Collision with buffer stops
HET 10	Passenger train collision with road vehicle on level crossings
HET 11	Non-passenger train collision with road vehicle on level crossings
HET 12	Derailment of passenger train
HET 13	Derailment of non-passenger train
	Movement accidents
HEM 1	Evacuation following stopped train
HEM 2	Passenger falls from train in running
HEM 3	Passenger struck while leaning out of train
HEM 4	Passenger struck by object through train window
HEM 5	Train door closes on passenger
HEM 6	Passenger falls between train and platform
HEM 7	Passenger falls out of train onto track at station
HEM 8	Passenger falls from platform and struck by train
HEM 9	Passenger fall/injured while boarding or alighting train
HEM 10	Passenger hit by object while on platform
HEM 11	Passenger struck while crossing track at station on crossing
HEM 12	MOP trespasser struck while crossing track at station

Table 1: List of hazardous events modelled within the SRM (Cont)

Haz Event Nunber	Hazardous Event Description
	Non-movement accidents
HEN 1	Lineside fire (other than in station)
HEN 2	Lineside fire in station
HEN 3	Fire in station
HEN 4	Lineside explosion
HEN 5	Explosion at Station
HEN 6	Passenger exposure to hazardous substances leakage
HEN 7	Passenger exposure to hazardous substances at station
HEN 8	Passenger exposed to electrical arcing at station
HEN 9	Passenger electric shock at station (OHL)
HEN 10	Passenger electric shock at station (Conductor rail)
HEN 11	Passenger electric shock at station (Non-traction supplies)
HEN 12	Passengers at station exposed to smoke or fumes
HEN 13	Passenger falls from platform onto to track (no train present)
HEN 14	Passenger slips, trips and falls
HEN 15	Passenger falls from overbridge at station
HEN 16	Passenger fall during evacuation at station
HEN 17	Passenger crushing caused by overcrowding at station
HEN 18	Train Crew exposure to hazardous substances

Hazardous event	Hazardous event description	Main precursor category	Risk without TPWS (eq.fats./ yr.)	Risk with TPWS (eq.fats./ yr.)	Risk reduction (eq.fats./ yr.)	% Risk reduction
HET-01	Collision between two passenger trains	Signal Passed at danger (SPAD)	2.65	0.99	1.66	62.6%
HET-02P	Collision between a passenger train and non-passenger train resulting from a passenger train SPAD	SPAD	0.47	0.18	0.29	61.3%
HET-02NP	Collision between a non-passenger train and passenger train resulting from a non- passenger train	SPAD	0.61	0.28	0.33	53.8%
HET-03	Collision between two non-passenger trains	SPAD	0.06	0.03	0.03	52.3%
HET-09	Collision with buffer stops	Misjudge braking	0.33	0.27	0.06	17.7%
HET-12	Derailment of passenger trains	SPAD	0.224	0.0923	0.13	58.8%
		Overspeed	0.037	0.012	0.03	68.6%
HET-13	Derailment of non-passenger trains	SPAD	0.129	0.053	0.08	58.9%
		Overspeed	0.033	0.010	0.02	68.5%
Total			4.54	1.92	2.62	57.7%

Table 3: Risk reduction resulting from TPWS fitment

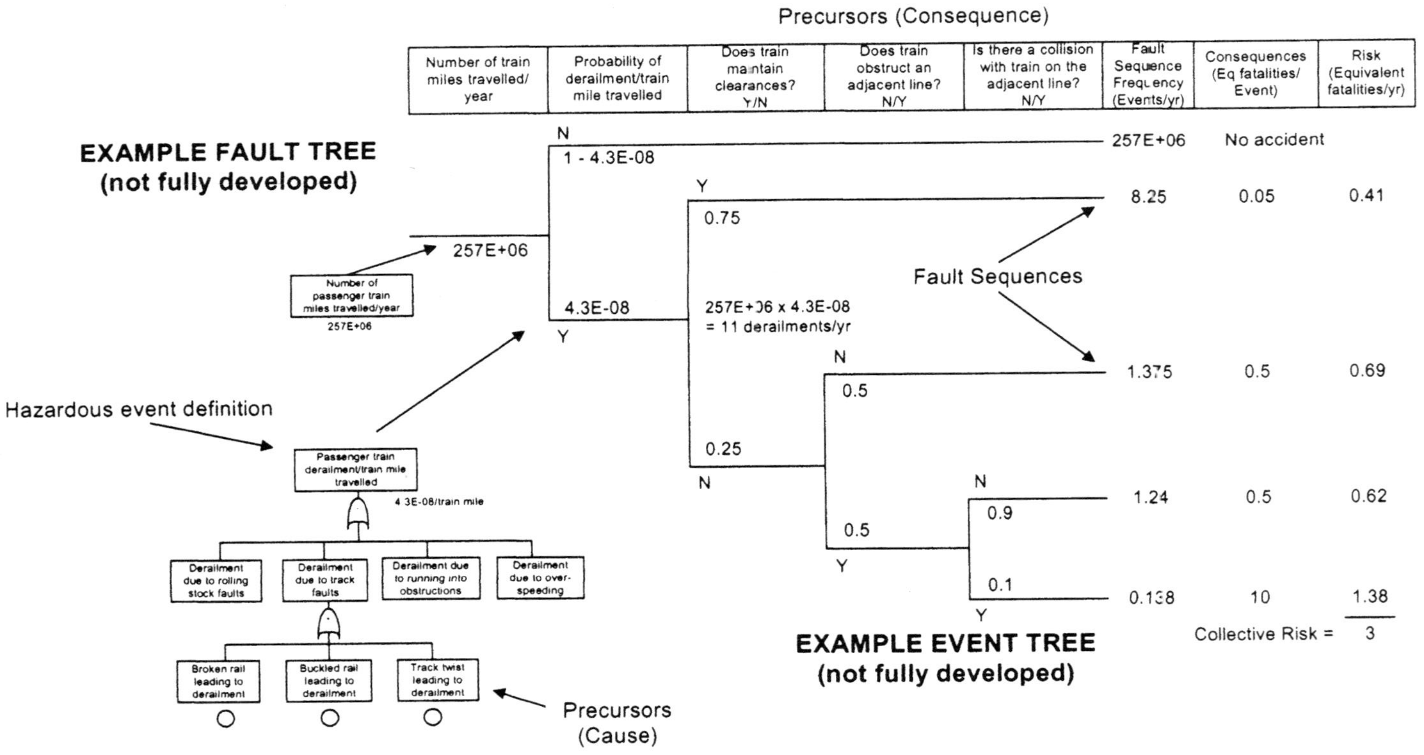

Figure 2: The relationship between fault tree and event tree analysis

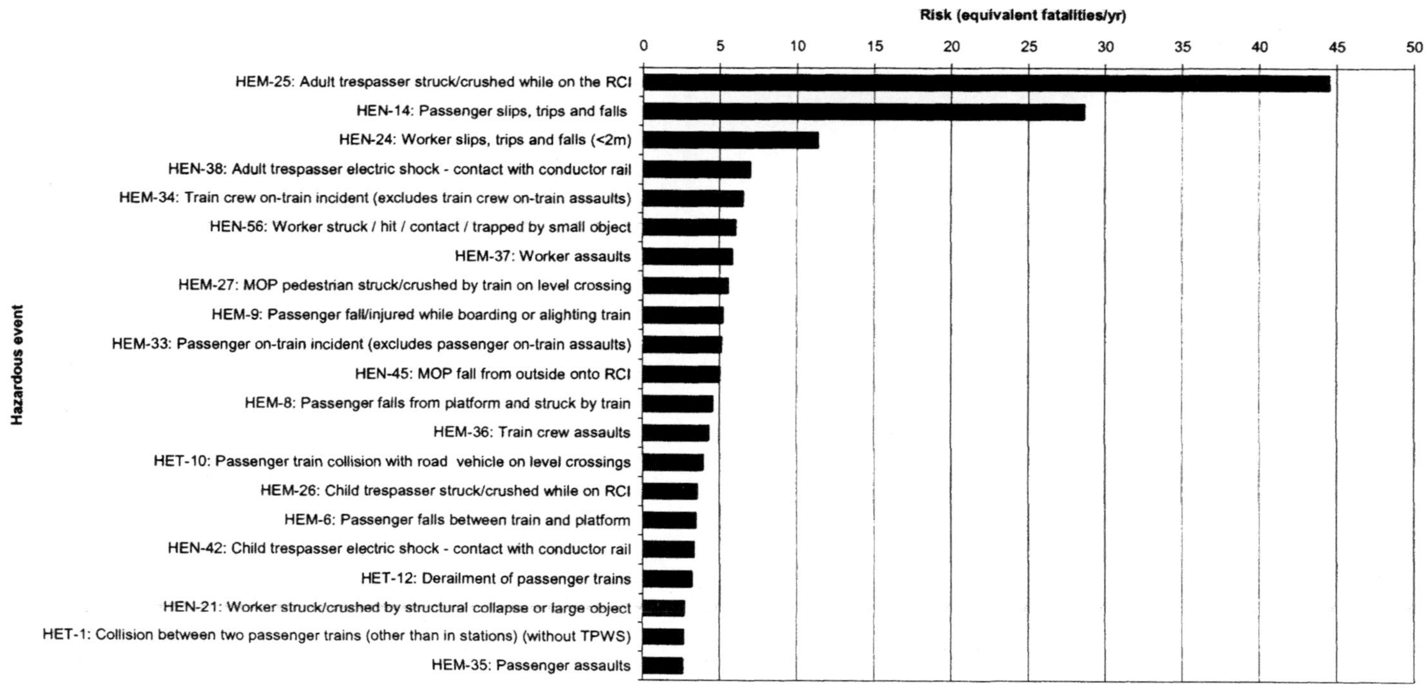

Figure 3: Risk Profile for hazardous events > 2 equivalent fatalities per year (without TPWS)

Safety Decision Making for the Railway

Andrew Sharpe
Head of Research and Development
Rail Safety and Standards Board
andrew.sharpe@rssb.co.uk

1 The Rail Safety and Standards Board

The Rail Safety and Standards Board (RSSB) was established on 1 April 2003 as a not-for-profit company owned by the railway industry. Our role is to provide leadership in the development of the long term safety strategy and policy for the UK railway. The company is limited by guarantee and has a members' council, a board and an advisory committee. It is independent of any single railway company and of their commercial interests.

One of RSSB's functions is to manage a programme of research on behalf of the industry and its wider stakeholders. Research into a wide range of issues is carried out with the aim of improving safety and reducing the cost of delivering a safe railway. The programme includes engineering research on subjects as diverse as the wheel-rail interface, train protection systems, and the effects of climate change on the infrastructure. It includes research into human factors and public behaviour, such as trespass and vandalism; and addresses management issues such as competence and safety culture.

In addition, the research programme is investigating policy issues in safety management and the governance of the railway industry. It is these policy issues that are the subject of this paper.

2 Safety Policy

How safe should the railway be? What criteria should we use when making decisions about safety – and what processes? The railway has many stakeholders (including passengers, freight customers, the general public, political representatives, and the industry itself): what levels of safety do they believe the railway should deliver? And do these stakeholders understand and support the way safety is managed?

These questions are central to the strategic management of the industry. They are not 'only' about safety: they have major implications for other industry outputs (capacity, performance, *etc*) and, crucially, for the industry's costs. Answering these questions is therefore vital to the future of the industry. Unfortunately, we do not have answers – not, at least, answers that attract a useful level of consensus. Industry and Government must therefore strive to address these questions in an intellectually robust and practical way, and prepare to implement changes that might be implied by the results.

This paper is a contribution to the efforts required in this area. The paper:

- Summarises accepted quantitative criteria for safety decisions
- Describes differences between these criteria and some current risk control actions
- Identifies tensions between different stakeholder perspectives
- Presents an outline of work currently under way or planned to address these issues.

3 Background: the ALARP Principle

The Health and Safety at Work Act 1974 (HSWA) [1], [2], [3] establishes in statute law the concept of reasonable practicability. Safety risks must be reduced *so far as is reasonably practicable*: if measures are available to reduce risk, they must be implemented unless the costs of doing so would be grossly disproportionate to the benefit obtained. This is often known as the ALARP principle – safety risk should be *As Low As Reasonably Practicable*. The principle is a means by which regulators seek to achieve two objectives:

- To ensure that organisations strive to improve safety, whilst taking into account the need for activities to continue in an economic manner.
- To ensure that judgements about safety are made in a similar way across all organisations and activities regulated by the HSWA.

The HSWA, and therefore the ALARP principle, apply by law to the management of safety on Britain's railways.

4 The Quantitative Approach

An organisation responsible for managing safety risk has to determine whether risk is already as low as reasonably practicable; conversely, it has to determine whether potential risk reductions are reasonably practicable to implement. This decision can be made in different ways to match different circumstances. For example, compliance with an accepted code of practice is sometimes regarded as a good indicator of risk being ALARP; or it may be that professional judgement is the best means of making a decision.

There are circumstances in which substantial investment is required to deliver an improvement – for example, when costly technology is available that will reduce the likelihood of a human error escalating into an accident. For the ALARP principle to be applied in such circumstances, it is necessary to have a quantitative method of comparing costs with safety benefits.

The method adopted in many industries is to use a Value of Preventing a Fatality (VPF). The VPF is the amount that an organisation will spend to reduce risk by a single fatality, and is used in cost benefit analysis (CBA) to assess reasonable practicability. The costs and benefits of a potential risk control are evaluated, and if the cost per life saved is less than or roughly equal to the VPF, the risk control is regarded as reasonably practicable and must therefore be implemented. The quantitative approach was formalised by the Health and Safety Executive (HSE) in its 1988 paper (updated in 1992) *The Tolerability of Risk from Nuclear Power Stations* [4] and its 1989 paper *Quantified Risk Assessment: its Input to Decision Making* [5]. Whilst the 1988 paper was developed for the nuclear industry, its principles have been applied widely.

5 The ALARP Principle in the Railway Industry

When CBA is used in the railway industry to determine reasonable practicability, the VPF employed is that defined, and updated annually, by RSSB on behalf of the industry in the annual Railway Group Safety Plan [6]. It is defined as follows:

- A default value of £1.30m, which is broadly consistent with values used, in principle, by Government in road safety. (In practice the amounts invested in road safety are less than this suggests, partly because the HSWA is not regarded as applying to the provision of public roadways.)
- A higher value of £3.64m, which is used when it is considered necessary to tip the balance further in favour of railway safety. For example, this higher value is used when the risk controls being evaluated are aimed at preventing or mitigating an accident that could involve multiple fatalities.

The use of two VPFs is an attempt to reflect what is referred to as 'societal concern' – broadly speaking, this is the perceived tendency for society as a whole to be more concerned about major accidents that lead to many fatalities, than about an equal or greater number of fatalities that occur in a succession of smaller accidents (*eg* train accidents *versus* road traffic accidents) [7].

What this approach means in practice can be illustrated with a simplified example:

- Suppose a new technology can be implemented at a cost of £100m, and risk assessment shows that over time it would be expected to save 50 lives in train accidents. A train accident can cause multiple fatalities, so the higher VPF would be applied. The cost per fatality avoided would in this case be £2m, which is less than the VPF of £3.64m, so the technology would be judged reasonably practicable and would be implemented.

- On the other hand, if risk assessment showed that the expected number of lives saved was five, the cost per fatality avoided would be £20m. This is much greater than the VPF, so the technology would be judged not to be reasonably practicable. It might still be implemented for other reasons, but the straightforward application of the test of reasonable practicability to the investment would not require its implementation.

6 Recent Developments

The railway industry's interpretation of 'reasonable practicability' as applied to investment decisions, described above, has been used across the industry for some years. However, it is not without its critics – some suggest, for example, that the railway industry's interpretation of the ALARP principle places too much emphasis on CBA and the quantitative comparison of risks *versus* costs. Moreover, recent developments have raised serious issues about the tenability of this approach.

6a Investment in Costly Safety Improvements

In the early 1990s, Government determined that an Automated Train Protection system (ATP) should not be installed across the bulk of the railway system, largely on the grounds of cost. ATP would prevent some accidents due to signals passed at danger (SPADs) and overspeeding, but these accidents are in any case rare so the total number of lives saved would be relatively small. CBA showed that the cost per life saved would be in the region of £14m – many times the VPF.

In response to this decision, the railway industry (led by Railtrack) developed the Train Protection and Warning System (TPWS) as a lower-cost alternative for installation at higher risk locations. Government determined that TPWS should be installed at a wide range of signals across the network, and the industry is completing its implementation. Estimates of the cost per fatality avoided are in the region of £8m to £10m. This cost, whilst lower than that for a full ATP system, is substantially greater than the agreed VPF of £3.64m.

Similar Government regulation requires the Train Operating Companies to replace some categories of older rolling stock ('unmodified Mark 1'), also at a cost much higher than would be justified by the agreed VPF.

These measures, it is argued, are justified to meet increasingly high public expectations for railway safety, to avoid 'catastrophic' risks, and to match good practice elsewhere. They imply a VPF of around £10m, and the industry has therefore adopted a comparable VPF to justify measures addressing certain other aspects of catastrophic risk. Examples are standards for on-train data recorders and for enhanced emergency braking.

6b HSE Guidance

In 2001 the HSE published a discussion document entitled *Reducing Risks, Protecting People* (known as R2P2) [8]. This document is the successor to the 1992 paper *The Tolerability of Risks from Nuclear Power Stations*. R2P2 develops and articulates the ALARP principle, and emphasises the need to take into account societal concerns about safety and health risks. It gives examples of types of risk to which, it is believed, societal concerns are most likely to apply:

- Risks that could lead to 'catastrophic' consequences – usually meaning those that could lead to multiple fatalities
- Risks where the consequences may be irreversible, *eg* the release of genetically modified organisms
- Risks that lead to inequalities because they affect some people more than others (*eg* the siting of a chemical plant or a waste disposal facility)
- Risks that could pose a threat to future generations, such as toxic waste.

For the railways, the most significant of these categories is the first. Railways are a safe mode of transport, but they do give rise to the potential for accidents that kill many people at once – such as Ladbroke Grove (31 deaths in 1999) and Clapham (35 deaths in 1988).

R2P2 is written as an explanation of the approach taken by the HSE to exercising its own regulation and enforcement responsibilities; it is not presented explicitly as guidance for industry. However, it is clear that the HSE expects industry to work in a way that is consistent with the philosophy developed in R2P2, reinforcing as it does the requirement to devote a special effort – and a high level of resources – to controlling risks with potentially catastrophic consequences. R2P2 provides some indications as to how societal concern might be evaluated and addressed, but does not provide comprehensive guidance on these subjects. RSSB is currently drawing together its own approach, for application in the development of industry-wide standards [9].

6c Public Consultation Regarding 'Willingness to Pay'

In 1998 and in 2000 Professor Michael Jones-Lee *et al* were commissioned by Government to carry out a study addressing the tolerability of certain kinds of safety risk [10], [11], [12]. The basis of Jones-Lee's approach was the concept of 'willingness to pay' (WTP) – *ie* the research attempted to determine how much money people were prepared to see devoted to reducing risks. The specific risks addressed in the study were train accidents, road traffic accidents, domestic fires, and fires in public places.

Jones-Lee concluded that the willingness of the public to pay for risk reductions was similar across these four types of risk; there was no evidence for a substantially greater willingness to pay for reduction in train accident risks on the grounds that they can lead to multi-fatality accidents. For the governance of safety risks on the railway, the main quantitative conclusions from Jones-Lee's work are (with figures adjusted for the passage of time):

- The VPF of £1.30m is justifiable on grounds of public preferences.
- The VPF of £3.64m is hard to justify on similar grounds, and must instead be justified in terms of avoiding reputational damage (to those responsible for the railway, including Government, regulators, and railway companies) and perhaps in terms of avoiding a public loss of confidence in the railway.
- The even higher implicit VPF of around £10m for TPWS or other risk controls is even harder to justify on grounds of safety.

However, Jones-Lee's reports include quotations from focus groups, in which members of the public expressed a wide range of views about recent railway accidents – including some that were highly critical of those in charge of the railway – suggesting that, to those members of the public at least, the risks leading to train accidents were in some sense 'unacceptable'.

7 The Diversity of Stakeholder Views

The Jones-Lee reports illustrate a widespread problem: not only do different stakeholders have different views about railway safety (and how much should be spent to improve safety), but sometimes a single individual can hold separate views that – when analysed in the light of factual information – appear to be mutually inconsistent. Put simply, it appears that the same people can at times believe that:

- risk is low and costs are high, so no more money should be made available;
- train crashes are avoidable and must not be allowed – those responsible are negligent.

These two views are not necessarily contradictory, but there can be circumstances in which they tend to work against each other – for example, where risk can be reduced by the introduction of a costly technology.

Media treatment of railway safety tends to exacerbate the difficulty of navigating this confusing set of issues and opinions. For example, shortly after the crash at Potters Bar in May 2002 *The Observer* [13] commented:

> *'The concentration of so many railway disasters – 17 in just 14 years – highlights the feebleness of Britain's current rail structure.'*

The implication is that 17 fatal train accidents in 14 years (an average of 1.2 accidents per year) constitutes a poor and deteriorating safety record. However, this rate of accidents in fact corresponds to a dramatic reduction in the number of fatal train accidents over the last few decades (Figure 1).

In such circumstances, it is not surprising that stakeholders, including the various branches of government, tend to emphasise different aspects of the problem. Some stress the need for the approach to safety to be rational, affordable, and pragmatic; others highlight the unacceptability of accidents and the continuing need for improvements in safety.

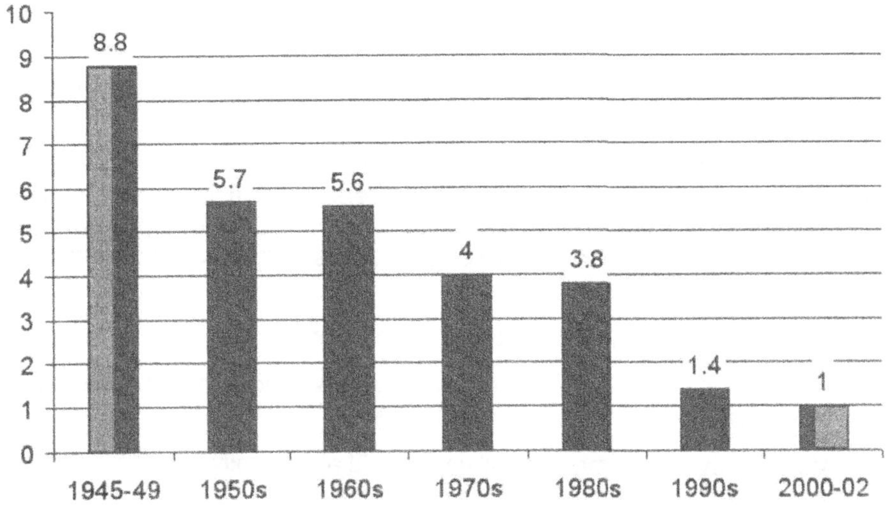

Figure 1: Fatal train accidents per year

The industry must take these views into account whilst managing the railway day-to-day and making plans for the future. People who work in the industry can be expected to show a degree of confusion and frustration. It is not uncommon to hear views expressed along the following lines:

- We stick to the rules on safety investment, and are criticised for it.
- We keep improving safety – will they never be satisfied?
- We are in a trap – the drive for safety and the shortage of money are irreconcilable.

The great divergence of views and of emphasis is indeed a trap, since it leads to confusion regarding the objectives set by society for the railway and undermines the claim that the regulation and management of safety have democratic support. It can lead to major decisions being made without a full understanding of their safety and economic implications.

8 Fundamental Issues

The foregoing discussion reveals a range of fundamental issues that have yet to be resolved. These include:

- How safe does the railway need to be?
- What level of safety do stakeholders expect the railway to deliver?

- Is it a question of 'level of safety', or is it more 'how safety is managed'?
- Should the industry and government always strive for lower safety risk?
- A great deal of money is spent in the name of safety. Is it spent wisely? Does it improve safety?
- There are lives and £billions at stake; are there better ways to make decisions?

The fact that we do not have answers to these key questions – not, at least, answers that are agreed by all stakeholders – is an extremely serious problem for the industry and for Government. Conversely, if we can generate a consensus around a set of responses to these questions, the rewards will be enormous:

- The avoidance of disproportionate expenditure.
- Better communications with stakeholders, including the railway's customers and political decision-makers.
- More clarity about what the railway should aim to deliver.
- Improved reputation for the industry and the relevant parts of Government.

Large though the rewards are, the problem is unlikely to be amenable to an instant solution. What is needed is a combination of the following:

(a) New thinking that allows stakeholders to escape from the sterility of opposing views from different perspectives.

(b) Research to structure and inform the debate about what the criteria and processes should be for safety decisions.

(c) Timely but deliberate action to engage stakeholders in this debate.

(d) Agreement amongst stakeholders as to the criteria and processes needed.

(e) Management arrangements for implementation and maintenance of the criteria and processes.

Items (a) and (b) are already under way through the RSSB's research programme, and are discussed in Sections 9 and 10 below; items (c), (d), and (e) are the subject of current work within the industry, facilitated by RSSB.

9 The Ethical Dimension

Risk is an ethical issue. Who is likely to be harmed, in what kind of accidents; how much does it cost to reduce the risk, and who pays – all these problems are essentially a question of how to manage a public service in a way that is ethically and politically acceptable.

RSSB's research in the ethical arena began in the summer of 2002 with work by the present author, which attempted to distinguish between attitudes to railway safety as seen from two different ethical perspectives: consequence-based and rule-based. This led to two research projects, aimed at injecting some new thinking by taking an explicitly ethical approach:

- *Railway Safety and the Ethics of the Tolerability of Risk* – a study by Jonathan Wolff, Professor of Philosophy at University College, London [14], [15].
- *Ethical Basis of Rail Safety Decisions* – a study by Chris Elliott of Pitchill Consulting and Tony Taig of TTAC [16].

These two studies have led to (largely) similar and complementary conclusions. Some of the highlights are:

- The established approach (including CBA) is for many aspects of safety risk a reasonable way of allocating resources. However, it does not answer all concerns about safety investment decisions.
- The 'willingness to pay' approach to establishing decision criteria assumes that it is only the 'risk *versus* cost' equation that matters. In practice, people have other concerns as well. For example, there is a question of involvement in process – have the kind of people affected by risk been involved in making decisions about it? Is it really the level of risk that worries people, or is it concern about the competence of the industry in managing safety? Is it the perceived motivation for decisions (whether taken by industry or by Government) that concerns people?
- People see railway safety from two perspectives – consumer and citizen. As consumers – *ie* as passengers – people generally do not believe that the railway is unsafe. However as citizens, people may be concerned about how the railway is managed, particularly when accidents appear to be the result of incompetence.
- This distinction appears to reflect an underlying difference between attitudes to risk based on *fear* and on *blame*. A person's fear of a risk is likely to be determined by his assessment of the hazard and probability; whereas the blame he attributes (to individuals, companies or society for the existence of a risk) is far more likely to be influenced by his belief about the morality of the process which causes or sustains the hazard. Low fear of a risk is therefore compatible with high blame for its creation.
- There appears to be desire for railways to reflect the 'state of the art'. For example, some people may believe that in the 21st Century trains simply should not go through red signals – whatever the 'risk *versus* cost' equation may be.
- Whilst all these ideas are a plausible attempt to reflect public views, it is

at present hard to know what the public think – the industry and Government need to obtain a better and more subtle understanding of the state of public opinion.

- Conversely, most members of the public – and many who are closely involved as stakeholders in the industry – lack knowledge of the facts and practicalities of railway safety. The industry must communicate better.
- As well as the formal structure of regulation, ideas from the field of corporate social responsibility tell us that any industry or company also has to have an 'informal licence to operate' – *ie* the public has to be broadly supportive of what that industry or company is doing. Arguably, the railway industry has come close to losing this informal licence at times in the recent past.
- It could be helpful to think in terms of some sort of compact or accord between the citizen, the state and the industry that defines much more clearly what the railway is expected to deliver in terms of safety, and other desirable outcomes such as capacity, reliability, punctuality, *etc*.

From this work and from subsequent discussions it is becoming clear that an important focus for research is *communication with* and *involvement of* the general public in how decisions are made. There is general agreement that people's preferences and concerns ought to be factored into these decisions, although there are differences of view on how and to what extent this should be done. Moreover, people have a right to be informed before expressing their preferences and concerns, although at present there is widespread lack of factual knowledge, apparently leading to many uninformed opinions and prejudices. Improving communication about safety is an important aim.

It is possible that the framework for decision making in the future will differ from the current VPF/CBA approach in terms of:

- Factoring in a richer range of concerns and preferences than are expressible via the 'Willingness to Pay' approach.
- Conducting the dialogue in a more open way, rather than depending on industry and government framing the issues, asking the questions, interpreting the answers, and implying that the result is a statement of the public's preferences.
- Improving communication both ways to enable industry and government to understand people's concerns and preferences once they have an unbiased understanding of the issues.

10 The Research Agenda

The scale of the problem, and the sensitivity and importance of the issues, makes it necessary to explore the way forward through a range of research projects. The broad objective of research in this area is to improve decision making – and the framework in which it operates – so as to improve the degree to which stakeholders support both the decisions and the way they have been made. The research work defined in this area addresses six main issues (figure 2).

Figure 2: Structure of Current Research

Fortunately, the railway is not alone in confronting these issues. For example, much work has been done on similar questions of managing health, safety, and environmental risk in the oil industry. The challenges facing the railway are not identical, but there are enough similarities to make cross-sectoral learning a realistic prospect [17].

11 Implementation

All of this research is necessary, and most of it is long overdue; but none of it will be of any use if the results cannot be implemented. RSSB is taking the lead in the steps necessary to progress towards implementation. These include early engagement of HM Government and the industry, and a structure for facilitating debate and agreement. RSSB will also seek to determine the steering and support functions needed to establish and maintain any new safety decision-making processes or criteria.

The entire exercise depends heavily on the engagement of stakeholders and on obtaining their views – delegates to this symposium in particular are invited to offer their views and provide their input to the debate about safety decision making for the railway.

12 References

1 *Health and Safety at Work etc. Act, 1974 (Chapter 37);* HMSO, 1991

2 *Management of Health and Safety at Work: Approved Code of Practice and Guidance;* HSE Books, 1999

3 *A Guide to Risk Assessment Requirements: Common Provisions in Health and Safety law;* HSE, 1996

4 *The Tolerability of Risk from Nuclear Power Stations* HSE 1988, revised 1992

5 *Quantified Risk Assessment: its Input to Decisions Making* HSE 1989

6 *Railway Group Safety Plan 2003-2004* RSSB, 2003

7 *Incorporating Public Perceptions of Risk into Safety Investment Decision Making* C Marris *et al;* for British Railways Board, November 1993

8 *Reducing Risks, Protecting People* Health and Safety Executive, 2001

9 *Consultation on the Railway Group Standards Code* RSSB, August 2003

10 *Valuation of Benefits of Health and Safety Control* Michael Jones-Lee *et al;* HSE, 2000

11 *Valuation of Benefits of Health and Safety Control: Follow-Up Study* Michael Jones-Lee *et al;* HSE, 2000

12 *ATP, SPADRAM, and the Cost-Benefit Analysis of Rail Safety* Michael Jones-Lee; Commissioned by Railtrack, June 2000

13 *The Observer*, 12 May 2002

14 *Railway Safety and the Ethics of the Tolerability of Risk* Jonathan Wolff for Railway Safety, March 2003

15 *Fear and Loathing in St Pancras* Jonathan Wolff for Railway Safety, presented to the IMechE Railway Safety Research Seminar, 24 April 2003

16 *Ethical Basis of Rail Safety Decisions* Chris Elliott and Tony Taig for Railway Safety, July 2003

17 *Perception vs Analysis: How to Handle Risk* Malcolm Brinded; Royal Academy of Engineering, Lloyd's Register Lecture, March 2000

SAFETY INTEGRITY LEVELS

Methods of Determining Safety Integrity Level (SIL) Requirements - Pros and Cons

W G Gulland
4-sight Consulting

1 Introduction

The concept of safety integrity levels (SILs) was introduced during the development of BS EN 61508 (BSI 2002) as a measure of the quality or dependability of a system which has a safety function – a measure of the confidence with which the system can be expected to perform that function. It is also used in BS IEC 61511(BSI 2003), the process sector specific application of BS EN 61508.

This paper discusses the application of 2 popular methods of determining SIL requirements – risk graph methods and layer of protection analysis (LOPA) – to process industry installations. It identifies some of the advantages of both methods, but also outlines some limitations, particularly of the risk graph method. It suggests criteria for identifying the situations where the use of these methods is appropriate.

2 Definitions of SILs

The standards recognise that safety functions can be required to operate in quite different ways. In particular they recognise that many such functions are only called upon at a low frequency / have a low demand rate. Consider a car; examples of such functions are:

- Anti-lock braking (ABS). (It depends on the driver, of course!).
- Secondary restraint system (SRS) (air bags).

On the other hand there are functions which are in frequent or continuous use; examples of such functions are:

- Normal braking
- Steering

The fundamental question is how frequently will failures of either type of function lead to accidents. The answer is different for the 2 types:

- For functions with a low demand rate, the accident rate is a combination of 2 parameters – i) the frequency of demands, and ii) the probability the function fails on demand (PFD). In this case, therefore, the appropriate measure of performance of the function is PFD, or its reciprocal, Risk Reduction Factor (RRF).

- For functions which have a high demand rate or operate continuously, the accident rate is the failure rate, λ, which is the appropriate measure of performance. An alternative measure is mean time to failure (MTTF) of the function. Provided failures are exponentially distributed, MTTF is the reciprocal of λ.

These performance measures are, of course, related. At its simplest, provided the function can be proof-tested at a frequency which is greater than the demand rate, the relationship can be expressed as:

$$PFD = \lambda T/2 \quad \text{or} \quad = T/(2 \times MTTF), \text{ or}$$
$$RRF = 2/(\lambda T) \quad \text{or} \quad = (2 \times MTTF)/T$$

where T is the proof-test interval. (Note that to significantly reduce the accident rate below the failure rate of the function, the test frequency, 1/T, should be at least 2 and preferably ≥ 5 times the demand frequency.) They are, however, different quantities. PFD is a probability – dimensionless; λ is a rate – dimension t^{-1}. The standards, however, use the same term – SIL – for both these measures, with the following definitions:

Table 1 - Definitions of SILs for Low Demand Mode from BS EN 61508

SIL	Range of Average PFD	Range of RRF[1]
4	$10^{-5} \leq PFD < 10^{-4}$	$100{,}000 \geq RRF > 10{,}000$
3	$10^{-4} \leq PFD < 10^{-3}$	$10{,}000 \geq RRF > 1{,}000$
2	$10^{-3} \leq PFD < 10^{-2}$	$1{,}000 \geq RRF > 100$
1	$10^{-2} \leq PFD < 10^{-1}$	$100 \geq RRF > 10$

Table 2 - Definitions of SILs for High Demand / Continuous Mode from BS EN 61508

SIL	Range of λ (failures per hour)	~ Range of MTTF (years)[2]
4	$10^{-9} \leq \lambda < 10^{-8}$	$100{,}000 \geq MTTF > 10{,}000$
3	$10^{-8} \leq \lambda < 10^{-7}$	$10{,}000 \geq MTTF > 1{,}000$
2	$10^{-7} \leq \lambda < 10^{-6}$	$1{,}000 \geq MTTF > 100$
1	$10^{-6} \leq \lambda < 10^{-5}$	$100 \geq MTTF > 10$

In low demand mode, SIL is a proxy for PFD; in high demand / continuous mode, SIL is a proxy for failure rate. (The boundary between low demand mode and high demand mode is in essence set in the standards at one demand per year. This is consistent with proof-test intervals of 3 to 6 months, which in many cases will be the shortest feasible interval.)

---

[1] This column is not part of the standards, but RRF is often a more tractable parameter than PFD.

[2] This column is not part of the standards, but the author has found these approximate MTTF values to be useful in the process industry sector, where time tends to be measured in years rather than hours.

Now consider a function which protects against 2 different hazards, one of which occurs at a rate of 1 every 2 weeks, or 25 times per year, i.e. a high demand rate, and the other at a rate of 1 in 10 years, i.e. a low demand rate. If the MTTF of the function is 50 years, it would qualify as achieving SIL1 for the high demand rate hazard. The high demands effectively proof-test the function against the low demand rate hazard. All else being equal, the effective SIL for the second hazard is given by:

$$PFD = 0.04/(2 \times 50) = 4 \times 10^{-4} \equiv SIL3$$

So what is the SIL achieved by the function? Clearly it is not unique, but depends on the hazard and in particular whether the demand rate for the hazard implies low or high demand mode.

In the first case, the achievable SIL is intrinsic to the equipment; in the second case, although the intrinsic quality of the equipment is important, the achievable SIL is also affected by the testing regime. This is important in the process industry sector, where achievable SILs are liable to be dominated by the reliability of field equipment – process measurement instruments and, particularly, final elements such as shutdown valves – which need to be regularly tested to achieve required SILs.

The differences between these definitions may be well understood by those who are dealing with the standards day-by-day, but are potentially confusing to those who only use them intermittently.

3 Some Methods of Determining SIL Requirements

BS EN 61508 offers 3 methods of determining SIL requirements:

- Quantitative method.
- Risk graph, described in the standard as a qualitative method.
- Hazardous event severity matrix, also described as a qualitative method.

BS IEC 61511 offers:

- Semi-quantitative method.
- Safety layer matrix method, described as a semi-qualitative method.
- Calibrated risk graph, described in the standard as a semi-qualitative method, but by some practitioners as a semi-quantitative method.
- Risk graph, described as a qualitative method.
- Layer of protection analysis (LOPA). (Although the standard does not assign this method a position on the qualitative / quantitative scale, it is weighted toward the quantitative end.)

Risk graphs and LOPA are popular methods for determining SIL requirements, particularly in the process industry sector. Their advantages and disadvantages and range of applicability are the main topic of this paper.

4 Risk Graph Methods

Risk graph methods are widely used for reasons outlined below. A typical risk graph is shown in Figure 1.

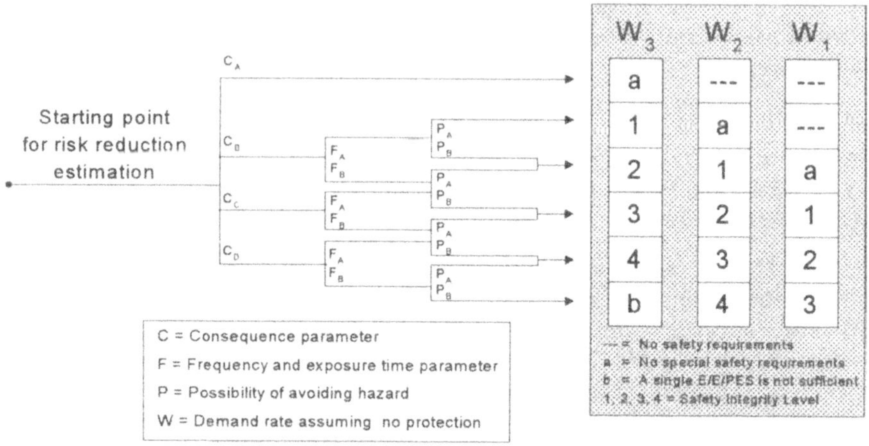

Figure 1 - Typical Risk Graph

The parameters of the risk graph can be given qualitative descriptions, e.g.:

$C_C \equiv$ death of several persons.

or quantitative descriptions, e.g.:

$C_C \equiv$ probable fatalities per event in range 0.1 to 1.0.

Table 3 - Typical Definitions of Risk Graph Parameters

Consequence		
C_A	Minor injury	
C_B	0.01 to 0.1 probable fatalities per event	
C_C	> 0.1 to 1.0 probable fatalities per event	
C_D	> 1 probable fatalities per event	
Exposure		
F_A	< 10% of time	
F_B	≥ 10% of time	
Avoidability / Unavoidability		
P_A	> 90% probability of avoiding hazard	< 10% probability hazard cannot be avoided
P_B	≤ 90% probability of avoiding hazard	≥ 10% probability hazard cannot be avoided
Demand Rate		
W_1	< 1 in 30 years	
W_2	1 in > 3 to 30 years	
W_3	1 in > 0.3 to 3 years	

The first definition begs the question "What does several mean?" In practice it is likely to be very difficult to assess SIL requirements unless there is a set of agreed definitions of the parameter values, almost inevitably in terms of quantitative ranges. These may or may not have been calibrated against the assessing organisation's risk criteria, but the method then becomes semi-quantitative (or is it semi-qualitative? It is certainly somewhere between the extremities of the qualitative / quantitative scale.)

Table 3 shows a typical set of definitions.

4.1 Benefits

Risk graph methods have the following advantages:

- They are semi-qualitative / semi-quantitative.
  - Precise hazard rates, consequences, and values for the other parameters of the method, are not required.
  - No specialist calculations or complex modelling is required.
  - They can be applied by people with a good "feel" for the application domain.
- They are normally applied as a team exercise, similar to HAZOP.
  - Individual bias can be avoided.
  - Understanding about hazards and risks is disseminated among team members (e.g. from design, operations, and maintenance).
  - Issues are flushed out which may not be apparent to an individual.
  - Planning and discipline are required.
- They do not require a detailed study of relatively minor hazards.
  - They can be used to assess many hazards relatively quickly.
  - They are useful as screening tools to identify:
    - hazards which need more detailed assessment
    - minor hazards which do not need additional protection

    so that capital and maintenance expenditures can be targeted where they are most effective, and lifecycle costs can be optimised.

4.2 The Problem of Range of Residual Risk

Consider the example: C_C, F_B, P_B, W_2 indicates a requirement for SIL3.

C_C ≡ > 0.1 to 1 probable fatalities per event
F_B ≡ ≥ 10% to 100% exposure
P_B ≡ ≥ 10% to 100% probability that the hazard cannot be avoided
W_2 ≡ 1 demand in > 3 to 30 years
SIL3 ≡ 10,000 ≥ RRF > 1,000

If all the parameters are at the geometric mean of their ranges:

Consequence = √(0.1 x 1.0) probable fatalities per event
= 0.32 probable fatalities per event
Exposure = √(10% x 100%) = 32%

Unavoidability = √(10% x 100%) = 32%
Demand rate = 1 in √(3 x 30) years
= 1 in ~10 years
RRF = √(1,000 x 10,000) = 3,200

(Note that geometric means are used because the scales of the risk graph parameters are essentially logarithmic.)

For the unprotected hazard:

Worst case risk = (1 x 100% x 100%) / 3 fatalities per year
= 1 fatality in ~3 years
Geometric mean risk = (0.32 x 32% x 32%) / 10 fatalities per year
= 1 fatality in ~300 years
Best case risk = (0.1 x 10% x 10%) / 30 fatalities per year
= 1 fatality in ~30,000 years

i.e. the unprotected risk has a range of 4 orders of magnitude.

With SIL3 protection:

Worst case residual risk = 1 fatality in (~3 x 1,000) years
= 1 fatality in ~3,000 years
Geometric mean residual risk = 1 fatality in (~300 x 3,200) years
= 1 fatality in ~1 million years
Best case residual risk = 1 fatality in (~30,000 x 10,000) years
= 1 fatality in ~300 million years

i.e. the residual risk with protection has a range of 5 orders of magnitude.

Figure 2 shows the principle, based on the mean case.

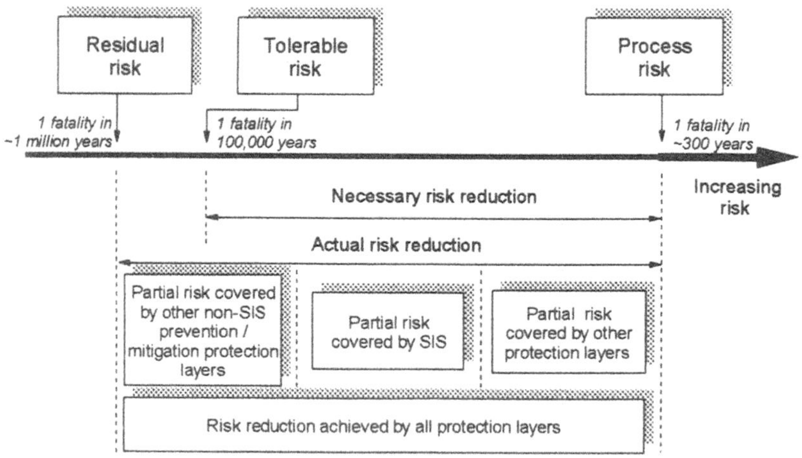

Figure 2 - Risk Reduction Model from BS IEC 61511

A reasonable target for this single hazard might be 1 fatality in 100,000 years. In the worst case we achieve less risk reduction than required by a factor of 30; in the mean case we achieve more risk reduction than required by a factor of 10; and in the best case we achieve more risk reduction than required by a factor of 3,000. In practice, of course, it is most unlikely that all the parameters will be at their extreme values, but on average the method must yield conservative results to avoid any significant probability that the required risk reduction is under-estimated.

Ways of managing the inherent uncertainty in the range of residual risk, to produce a conservative outcome, include:

- Calibrating the graph so that the mean residual risk is significantly below the target, as above.
- Selecting the parameter values cautiously, i.e. by tending to select the more onerous range whenever there is any uncertainty about which value is appropriate.
- Restricting the use of the method to situations where the mean residual risk from any single hazard is only a very small proportion of the overall total target risk. If there are a number of hazards protected by different systems or functions, the total mean residual risk from these hazards should only be a small proportion of the overall total target risk. It is then very likely that an under-estimate of the residual risk from one hazard will still be a small fraction of the overall target risk, and will be compensated by an over-estimate for another hazard when the risks are aggregated.

This conservatism may incur a substantial financial penalty, particularly if higher SIL requirements are assessed.

4.3 Use in the Process Industries

Risk graphs are popular in the process industries for the assessment of the variety of trip functions – high and low pressure, temperature, level and flow, etc – which are found in the average process plant. In this application domain, the benefits listed above are relevant, and the criterion that there are a number of functions whose risks can be aggregated is usually satisfied.

Figure 3 shows a typical function. The objective is to assess the SIL requirement of the instrumented over-pressure trip function (in the terminology of BS IEC 61511, a "safety instrumented function", or SIF, implemented by a "safety instrumented system", or SIS). One issue which arises immediately, when applying a typical risk graph in a case such as this, is how to account for the relief valve, which also protects the vessel from over-pressure. This is a common situation – a SIF backed up mechanical protection. The options are:

- Assume it ALWAYS works
- Assume it NEVER works
- Something in-between

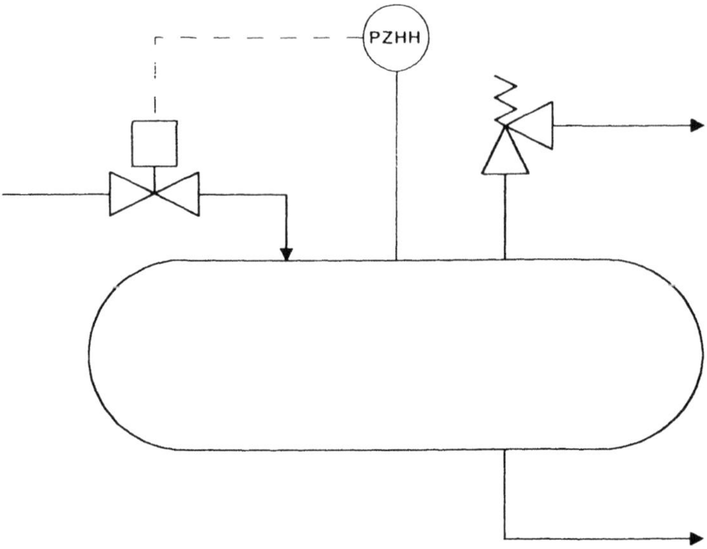

Figure 3 - High Pressure Trip Function

The first option was recommended in the UKOOA Guidelines (UKOOA 1999), but cannot be justified from failure rate data. The second option is liable to lead to an over-estimate of the required SIL, and to incur a cost penalty, so cannot be recommended.

See Table 4 for the guidance provided by the standards.

An approach which has been found to work, and which accords with the standards is:

1. Derive an overall risk reduction requirement (SIL) on the basis that there is no protection, i.e. before applying the SIF or any mechanical protection.
2. Take credit for the mechanical device, usually as equivalent to SIL2 for a relief valve (this is justified by available failure rate data, and is also supported by BS IEC 61511, Part 3, Annex F)
3. The required SIL for the SIF is the SIL determined in the first step minus 2 (or the equivalent SIL of the mechanical protection).

The advantages of this approach are:

- It produces results which are generally consistent with conventional practice.
- It does not assume that mechanical devices are either perfect or useless.
- It recognises that SIFs require a SIL whenever the overall requirement exceeds the equivalent SIL of the mechanical device (e.g. overall requirement = SIL3; relief valve ≡ SIL2; SIF requirement = SIL1).

Table 4 - Guidance from the Standards on Handling "Other Technology Safety Related Systems" with Risk Graphs

BS EN 61508	BS IEC 61511
"The purpose of the W factor is to estimate the frequency of the unwanted occurrence taking place without the addition of any safety-related systems (E/E/PE or other technology) but including any external risk reduction facilities." (Part 5, Annex D – A qualitative method: risk graph) *(A relief valve is clearly an "other technology safety-related device".)*	"W - The number of times per year that the hazardous event would occur in the absence of the safety instrumented function under consideration. This can be determined by considering all failures which can lead to the hazardous event and estimating the overall rate of occurrence. Other protection layers should be included in the consideration." (Part 3, Annex D – semi-qualitative, calibrated risk graph) and: "The purpose of the W factor is to estimate the frequency of the hazard taking place without the addition of the SIS." (Part 3, Annex D – semi-qualitative, calibrated risk graph) "The purpose of the W factor is to estimate the frequency of the unwanted occurrence taking place without the addition of any safety instrumented systems (E/E/PE or other technology) but including any external risk reduction facilities." (Part 3, Annex E – qualitative, risk graph)

4.4 Calibration for Process Plants

Before a risk graph can be calibrated, it must first be decided whether the basis will be:

- Individual risk (IR), usually of someone identified as the most exposed individual.
- Group risk of an exposed population group, such as the workers on the plant or the members of the public on a nearby housing estate.
- Some combination of these 2 types of risk.

4.4.1 Based on Group Risk

Consider the risk graph and definitions developed above as they might be applied to the group risk of the workers on a given plant. If we assume that on the plant there are 20 such functions, then, based on the geometric mean residual risk (1 in 1 million years), the total risk is 1 fatality in 50,000 years.

Compare this figure with published criteria for the acceptability of risks. The HSE have suggested that a risk of one 50 fatality event in 5,000 years is intolerable (HSE Books 2001). They also make reference, in the context of risks from major industrial installations, to "Major hazards aspects of the transport of dangerous substances" (HMSO 1991), and in particular to the F-N curves it contains (Figure 4).

The "50 fatality event in 5,000 years" criterion is on the "local scrutiny line", and we may therefore deduce that 1 fatality in 100 years should be regarded as intolerable, while 1 in 10,000 years is on the boundary of "broadly acceptable". Our target might therefore be "less than 1 fatality in 1,000 years". In this case the total risk from hazards protected by SIFs (1 in 50,000 years) represent 2% of the overall risk target, which probably allows more than adequately for other hazards for which SIFs are not relevant. We might therefore conclude that this risk graph is over-calibrated for the risk to the population group of workers on the plant. However, we might choose to retain this additional element of conservatism to further compensate for the inherent uncertainties of the method.

To calculate the average IR from this calibration, let us estimate that there is a total of 50 persons regularly exposed to the hazards (i.e. this is the total of all regular workers on all shifts). The risk of fatalities of 1 in 50,000 per year from hazards protected by SIFs is spread across this population, so the average IR is 1 in 2.5 million (4E-7) per year.

Comparing this IR with published criteria from R2P2 (HSE Books 2001):

- Intolerable = 1 in 1,000 per year (for workers)
- Broadly acceptable = 1 in 1 million per year

Our overall target for IR might therefore be "less than 1 in 50,000 (2E-5) per year" for all hazards, so that the total risk from hazards protected by SIFs again represents 2% of the target, so probably allows more than adequately for other hazards, and we might conclude that the graph is also over-calibrated for average individual risk to the workers.

The C and W parameter ranges are available to adjust the calibration. (The F and P parameters have only 2 ranges each, and F_A and P_A both imply reduction of risk by at least a factor of 10.) Typically, the ranges might be adjusted up or down by half an order of magnitude.

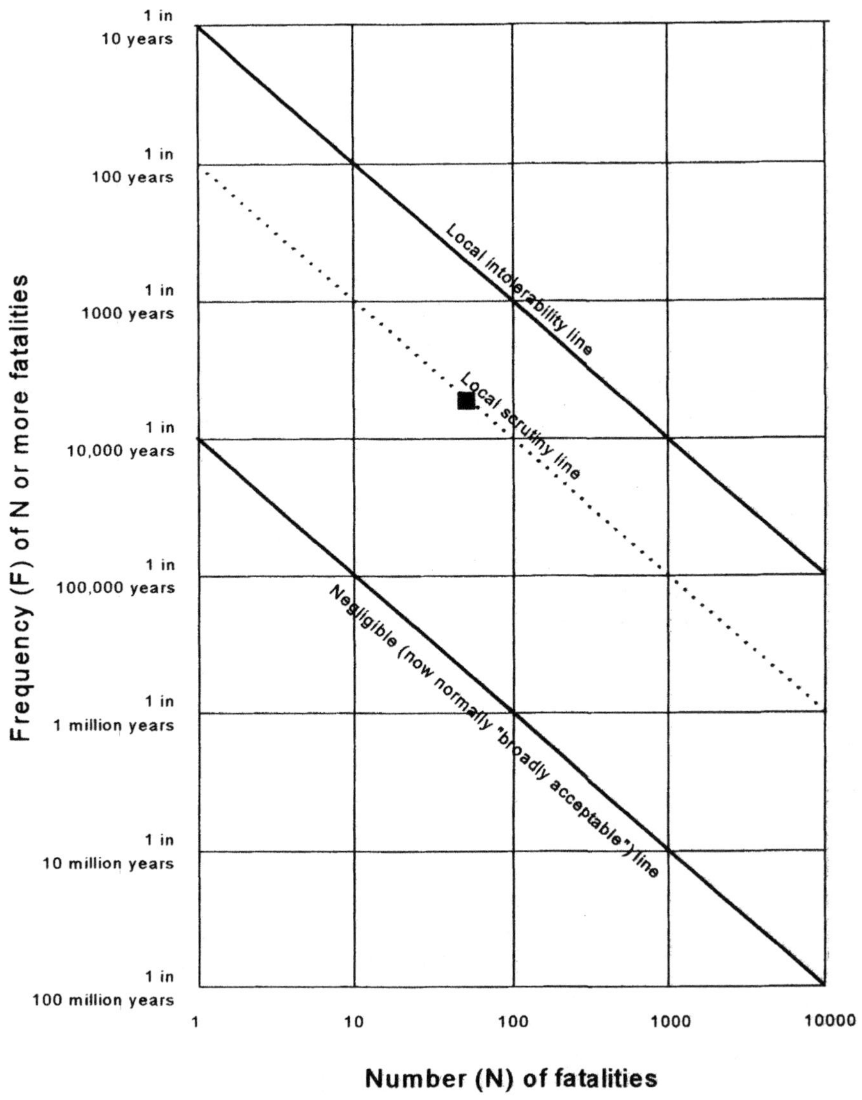

Figure 4 - F-N Curves from Major Hazards of Transport Study

The plant operating organisation may, of course, have its own risk criteria, which may be onerous than these criteria derived from R2P2 and the Major hazards of transport study.

4.4.2 Based on Individual Risk to Most Exposed Person

To calibrate a risk graph for IR of the most exposed person it is necessary to identify who that person is, at least in terms of his job and role on the plant. The values of the C parameter must be defined in terms of consequence to the individual, e.g.:

C_A ≡ Minor injury
C_B ≡ ~0.01 probability of death per event
C_C ≡ ~0.1 probability of death per event
C_D ≡ death almost certain

The values of the exposure parameter, F, must be defined in terms of the time he spends at work, e.g.:

F_A ≡ exposed for < 10% of time spent at work
F_B ≡ exposed for ≥ 10% of time spent at work

Recognising that this person only spends ~20% of his life at work, he is potentially at risk from only ~20% of the demands on the SIF. Thus, again using C_C, F_B, P_B and W_2:

C_C	≡ ~0.1 probability of death per event
F_B	≡ exposed for ≥ 10% of working week or year
P_B	≡ > 10% to 100% probability that the hazard cannot be avoided
W_2	≡ 1 demand in > 3 to 30 years
SIL3	≡ 1,000 ≥ RRF > 10,000

for the unprotected hazard:

Worst case risk	= 20% x (0.1 x 100% x 100%) / 3 probability of death per year
	= 1 in ~150 probability of death per year
Geometric mean risk	= 20% x (0.1 x 32% x 32%) / 10 probability of death per year
	= 1 in ~4,700 probability of death per year
Best case risk	= 20% x (0.1 x 10% x 10%) / 30 probability of death per year
	= 1 in ~150,000 probability of death per year

With SIL3 protection:

Worst case residual risk	= 1 in ~150,000 probability of death / year
Geometric mean residual risk	= 1 in ~15 million probability of death / year
Best case residual risk	= 1 in ~1.5 billion probability of death / year

If we estimate that this person is exposed to 10 hazards protected by SIFs (i.e. to half of the total of 20 assumed above), then, based on the geometric mean residual risk, his total risk of death from all of them is 1 in 1.5 million per year. This is 3.3% of our target of 1 in 50,000 per year IR for all hazards, which probably leaves more than adequate allowance for other hazards for which SIFs are not relevant. We might therefore conclude that this risk graph also is over-calibrated for the risks to our hypothetical most exposed individual, but we can choose to accept this additional element of conservatism. (Note that this is NOT the same risk graph as the one considered above for group risk, because, although we have retained the form, we have used a different set of definitions for the parameters.)

The above definitions of the C parameter values do not lend themselves to adjustment, so in this case only the W parameter ranges can be adjusted to re-calibrate the graph. We might for example change the W ranges to:

W_1 ≡ < 1 demand in 10 years
W_2 ≡ 1 demand in > 1 to 10 years
W_3 ≡ 1 demand in ≤ 1 year

4.5 Typical Results

As one would expect, there is wide variation from installation to installation in the numbers of functions which are assessed as requiring SIL ratings, but Table 5 shows figures which were assessed for a reasonably typical offshore gas platform.

Table 5 - Typical Results of SIL Assessment

SIL	Number of Functions	% of Total
4	0	0%
3	0	0%
2	1	0.3%
1	18	6.0%
None	281	93.7
Total	300	100%

Typically, there might be a single SIL3 requirement, while identification of SIL4 requirements is very rare.

These figures suggest that the assumptions made above to evaluate the calibration of the risk graphs are reasonable.

4.6 Discussion

The implications of the issues identified above are:

- Risk graphs are very useful but coarse tools for assessing SIL requirements. (It is inevitable that a method with 5 parameters – C, F, P, W and SIL – each with a range of an order of magnitude, will produce a result with a range of 5 orders of magnitude.)
- They must be calibrated on a conservative basis to avoid the danger that they under-estimate the unprotected risk and the amount of risk reduction / protection required.
- Their use is most appropriate when a number of functions protect against different hazards, which are themselves only a small proportion of the overall total hazards, so that it is very likely that under-estimates and over-estimates of residual risk will average out when they are aggregated. Only in these circumstances can the method be realistically described as providing a "suitable" and "sufficient", and therefore legal, risk assessment.

- Higher SIL requirements (SIL2+) incur significant capital costs (for redundancy and rigorous engineering requirements) and operating costs (for applying rigorous maintenance procedures to more equipment, and for proof-testing more equipment). They should therefore be re-assessed using a more refined method.

5 Layer of Protection Analysis (LOPA)

The LOPA method was developed by the American Institute of Chemical Engineers as a method of assessing the SIL requirements of SIFs (AIChemE 1993).

The method starts with a list of all the process hazards on the installation as identified by HAZOP or other hazard identification technique. The hazards are analysed in terms of:

- Consequence description ("Impact Event Description")
- Estimate of consequence severity ("Severity Level")
- Description of all causes which could lead to the Impact Event ("Initiating Causes")
- Estimate of frequency of all Initiating Causes ("Initiation Likelihood")

The Severity Level may be expressed in semi-quantitative terms, with target frequency ranges (see Table 6),

Table 6 - Example Definitions of Severity Levels and Mitigated Event Target Frequencies

Severity Level	Consequence	Target Mitigated Event Likelihood
Minor	Serious injury at worst	No specific requirement
Serious	Serious permanent injury or up to 3 fatalities	< 3E-6 per year, or 1 in > 330,000 years
Extensive	4 or 5 fatalities	< 2E-6 per year, or 1 in > 500,000 years
Catastrophic	> 5 fatalities	Use F-N curve

or it may be expressed as a specific quantitative estimate of harm, which can be referenced to F-N curves.

Similarly, the Initiation Likelihood may be expressed semi-quantitatively (see Table 7),

Table 7 - Example Definitions of Initiation Likelihood

Initiation Likelihood	Frequency Range
Low	< 1 in 10,000 years
Medium	1 in > 100 to 10,000 years
High	1 in ≤ 100 years

or it may be expressed as a specific quantitative estimate.

The strength of the method is that it recognises that in the process industries there are usually several layers of protection against an Initiating Cause leading to an Impact Event. Specifically, it identifies:

- General Process Design. There may, for example, be aspects of the design which reduce the probability of loss of containment, or of ignition if containment is lost, so reducing the probability of a fire or explosion event.
- Basic Process Control System (BPCS). Failure of a process control loop is likely to be one of the main Initiating Causes. However, there may be another independent control loop which could prevent the Impact Event, and so reduce the frequency of that event.
- Alarms. Provided there is an alarm which is independent of the BPCS, sufficient time for an operator to respond, and an effective action he can take (a "handle" he can "pull"), credit can be taken for alarms to reduce the probability of the Impact Event.
- Additional Mitigation, Restricted Access. Even if the Impact Event occurs, there may be limits on the occupation of the hazardous area (equivalent to the F parameter in the risk graph method), or effective means of escape from the hazardous area (equivalent to the P parameter in the risk graph method), which reduce the Severity Level of the event.
- Independent Protection Layers (IPLs). A number of criteria must be satisfied by an IPL, including RRF $\geq$ 100. Relief valves and bursting disks usually qualify.

Based on the Initiating Likelihood (frequency) and the PFDs of all the protection layers listed above, an Intermediate Event Likelihood (frequency) for the Impact Event and the Initiating Event can be calculated. The process must be completed for all Initiating Events, to determine a total Intermediate Event Likelihood for all Initiating Events. This can then be compared with the target Mitigated Event Likelihood (frequency). So far no credit has been taken of any SIF. The ratio:

(Intermediate Event Likelihood) / (Mitigated Event Likelihood)

gives the required RRF (or 1/PFD) of the SIF, and can be converted to a SIL.

5.1 Benefits

The LOPA method has the following advantages:

- It can be used semi-quantitatively or quantitatively.
  - Used semi-quantitatively it has many of the same advantages as risk graph methods.
  - Use quantitatively the logic of the analysis can still be developed as a team exercise, with the detail developed "off-line" by specialists.
- It explicitly accounts for risk mitigating factors, such as alarms and relief valves, which have to be incorporated as adjustments into risk graph methods

(e.g. by reducing the W value to take credit for alarms, by reducing the SIL to take credit for relief valves).

- A semi-quantitative analysis of a high SIL function can be promoted to a quantitative analysis without changing the format.

6 After-the-Event Protection

Some functions on process plants are invoked "after-the-event", i.e. after a loss of containment, even after a fire has started or an explosion has occurred. Fire and gas detection and emergency shutdown are the principal examples of such functions. Assessment of the required SILs of such functions presents specific problems:

- Because they operate after the event, there may already have been consequences which they can do nothing to prevent or mitigate. The initial consequences must be separated from the later consequences.
- The event may develop and escalate to a number of different eventual outcomes with a range of consequence severity, depending on a number of intermediate events. Analysis of the likelihood of each outcome is a specialist task, often based on event trees (Figure 5).

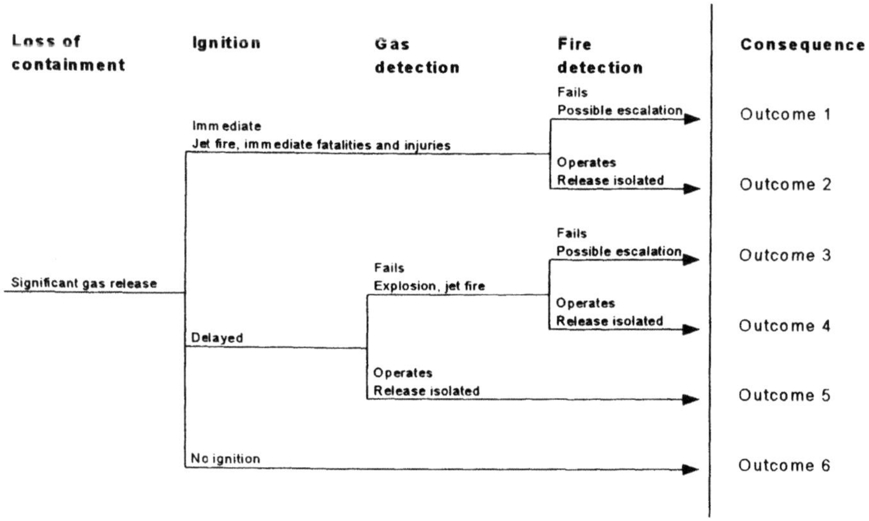

Figure 5 - Event Tree for After the Event Protection

The risk graph method does not lend itself at all well to this type of assessment:

- Demand rates would be expected to be very low, e.g. 1 in 1,000 to 10,000 years. This is off the scale of the risk graphs presented here, i.e. it implies a range 1 to 2 orders of magnitude lower than W_1.
- The range of outcomes from function to function may be very large, from a single injured person to major loss of life. Where large scale consequences are

possible, use of such a coarse tool as the risk graph method can hardly be considered "suitable" and "sufficient".

The LOPA method does not have these limitations, particularly if applied quantitatively.

7 Conclusions

To summarise, the relative advantages and disadvantages of these 2 methods are:

Risk Graph Methods	LOPA
Advantages: 1. Can be applied relatively rapidly to a large number of functions to eliminate those with little or no safety role, and highlight those with larger safety roles. 2. Can be performed as a team exercise involving a range of disciplines and expertise.	Advantages: 1. Can be used both as a relatively coarse filtering tool and for more precise analysis. 2. Can be performed as a team exercise, at least for a semi-quantitative assessment. 3. Facilitates the identification of all relevant risk mitigation measures, and taking credit for them in the assessment. 4. When used quantitatively, uncertainty about residual risk levels can be reduced, so that the assessment does not need to be so conservative. 5. Can be used to assess the requirements of after-the-event functions.
Disadvantages: 1. A coarse method, which is only appropriate to functions where the residual risk is very low compared to the target total risk. 2. The assessment has to be adjusted in various ways to take account of other risk mitigation measures such as alarms and mechanical protection devices.	Disadvantages: 1. Relatively slow compared to risk graph methods, even when used semi-quantitatively. 2. Not so easy to perform as a team exercise; makes heavier demands on team members' time, and not so visual.

Risk Graph Methods	**LOPA**
Disadvantages: 3. Does not lend itself to the assessment of after-the-event functions.	Disadvantages:

Both methods are useful, but care should be taken to select a method which is appropriate to the circumstances.

References

AIChemE (1993). Guidelines for Safe Automation of Chemical Processes, ISBN 0-8169-0554-1

BSI (2002). BS EN 61508, Functional safety of electrical / electronic / programmable electronic safety-related systems

BSI (2003). BS IEC 61511, Functional safety - Safety instrumented systems for the process industry sector

HMSO (1991). Major hazards aspects of the transport of dangerous substances, ISBN 0-11-885699-5

HSE Books (2001). Reducing risks, protecting people (R2P2), Clause 136, ISBN 0-7176-2151-0

UKOOA (1999). Guidelines for Instrument-Based Protective Systems, Issue No.2, Clause 4.4.3.

An Examination of the IEC 61508 Approach Towards Safety Integrity Levels and Modes of Operation of Safety Functions

S J Brown
Health & Safety Executive
Bootle, UK

Abstract

This paper examines the approach of IEC 61508 towards the determination of safety integrity levels. It examines the distinction made by the standard between 'low demand' and 'high demand or continuous' modes of operation and the possible changes being considered by the IEC working group.

1 Introduction

The overall aim of the international standard for electrical/electronic/ programmable electronic (E/E/PE) safety-related systems, IEC 61508, is to ensure that the likelihood of dangerous system failure is low enough, taking into account the hazards associated with the system's application. To this end, the standard requires the quantification of the likelihood of dangerous system failures that are due to the failure of hardware components. For systematic aspects, including any software components, the standard grades, in terms of safety integrity level (SIL), the effectiveness of a range of measures and techniques recommended for the avoidance of faults and control of failures. The SIL required for a particular application is determined by risk assessment. The SIL guides the selection of the measures and techniques and thereby gives the necessary confidence that the system will not fail due to systematic faults. The relationships between the risks involved, the quantified failure measure for hardware failures and the SILs for systematic aspects are defined by the standard according to whether the safety function under consideration operates in the 'low demand' or 'high demand or continuous' mode of operation. This paper gives the author's views of the approach of the standard and the underlying rationale.

2 IEC 61508 approach

The overall approach of IEC 61508 [IEC, 1998a] can be summarised as follows:

1) Specify the required performance of each of the safety functions of the safety-related system in terms of the risk reduction (for 'low demand

mode' safety functions) or the hazard rate (for 'high demand or continuous mode' safety functions), as appropriate, to achieve a tolerable level of residual risk. This is determined by a hazard and risk assessment that takes into account the behaviour of the associated equipment or process, the risk considered tolerable in the particular application, and the extent of any risk reduction provided by other safety-related systems, relevant to the hazard being considered.

2) Verify that the reliability of the hardware of the safety-related system, taking into account any periodic proof testing and automatic diagnostics, is such that the required risk reduction or hazard rate, as appropriate, is achieved.

3) Refer to IEC 61508-1, table 2 (for low demand mode operation) or table 3 (for continuous or high demand mode operation) to determine the safety integrity level (SIL). The SIL then guides the selection of the techniques used for the avoidance of systematic faults in both hardware and software, so that as the risk reduction increases, or the hazard rate decreases, there is a reduction in the likelihood that systematic failures (including those resulting from incorrect specification) will result in a hazard.

2.1 Low demand mode of operation

According to IEC 61508-4 [IEC, 1998c], 'low demand mode' safety functions are those where the 'frequency of demands on safety-related system is no greater than one per year and no greater than twice the proof test frequency'. The measure of safety performance of a demand mode safety function is the risk reduction factor, ΔR:

$$\Delta R = d / h \qquad (1)$$

where: h = target hazard rate to achieve tolerable risk

d = demand rate on the safety function

The demand rate is determined by the characteristics of the associated equipment under control (EUC) in combination with the reliability of the EUC control system. The hazard rate is determined by the risk considered to be tolerable in the particular application, taking into account the risk reduction provided by any other safety-related systems or risk reduction measures relevant to the hazard being considered.

The risk reduction factor gives the required average probability of failure on demand, $PFD_{avg(target)}$, of the safety function according to:

$$PFD_{avg(target)} = 1/\Delta R \qquad (2)$$

IEC 61508-1 table 2 (Figure 1), defines the relationship between $PFD_{avg(target)}$, and hence the risk reduction, and the SIL. The fundamental purpose of table 2 is to define this linkage between the safety performance of the safety function and the grading of the measures for avoidance of systematic failures. However, it should be recognised that the linkage is not established on a quantified basis. This is because, in general, it is not possible to quantify the impact of the measures for avoidance of systematic failures in terms of failure rates. Rather, the standard makes recommendations for systematic measures in relation to SIL on the basis of what is considered to be good practice for the associated level of risk reduction.

Safety integrity level	**Low demand mode of operation** (Average probability of failure to perform its design function on demand)
4	$\geq 10^{-5}$ to $< 10^{-4}$
3	$\geq 10^{-4}$ to $< 10^{-3}$
2	$\geq 10^{-3}$ to $< 10^{-2}$
1	$\geq 10^{-2}$ to $< 10^{-1}$

Figure 1 IEC 61508-1 table 2

2.1.1 Hardware safety requirement

From a hardware viewpoint, IEC 61508-2 [IEC, 2000a] requires that the reliability of the hardware, taking into account any periodic testing and automatic diagnostics, is such that random hardware failures do not cause dangerous failures of the safety-related system at a rate which would result in the target hazard rate being exceeded. This is achieved if the probability of failure on demand of the specific hardware design, $PFD_{avg(achieved)}$, does not exceed the target failure measure, $PFD_{avg(target)}$:

$$PFD_{avg(achieved)} \leqslant PFD_{avg(target)} \qquad (3)$$

$PFD_{avg(achieved)}$ is determined by the reliability of the hardware components, the extent of any hardware redundancy, the frequency of any periodic proof testing and the coverage of any automatic diagnostics. The standard requires an estimation of $PFD_{avg(achieved)}$ to confirm (3) for the specific hardware design. It allows the use of any of the established reliability modelling techniques including reliability block diagrams, Markov analysis, etc. IEC 61508-6 [IEC, 2000b] presents examples of how to calculate $PFD_{avg(achieved)}$ for a number of common architectures, based on the use of simplified equations which are considered reasonable approximations in the low demand region. These simplified equations are based on the approximation:

$$PFD_{avg(achieved)} \simeq \lambda.T/2 \quad (4)$$

Here, λ is the dangerous failure rate of the safety-related system and T is the time interval between proof tests. A further assumption of the simplified equations in IEC 61508-6 is that any faults in the safety-related system, revealed by the proof test, are repaired following the test.

2.1.2 Systematic safety requirement

The SIL given by table 2 for the required risk reduction forms the basis for the selection of measures and techniques for the avoidance of systematic failures according to the recommendations in IEC 61508-2 (for hardware) and IEC 61508-3 [IEC, 1998b] (for software).

2.2 High demand or continuous mode of operation

Safety functions are defined by IEC 61508-4 as operating in the 'high demand or continuous mode of operation' if the demand rate is greater than one per year or greater than twice the proof test frequency. In this case, as discussed below, the measure of the safety performance of the safety function is the limit of hazard rate, h, that achieves tolerable risk. The relationship between the quantified safety performance and the SIL is given in IEC 61508-1 table 3, see Figure 2.

Safety integrity level	**High demand or continuous mode of operation** (Probability of a dangerous failure per hour)
4	$\geq 10^{-9}$ to $< 10^{-8}$
3	$\geq 10^{-8}$ to $< 10^{-7}$
2	$\geq 10^{-7}$ to $< 10^{-6}$
1	$\geq 10^{-6}$ to $< 10^{-5}$

Figure 2 IEC 61508-1 table 3

Note that table 3 expresses the safety performance measure as the 'probability of a dangerous failure per hour'. This is because, at high demand rates (in relation to the proof test interval), the hazard rate for a demand mode safety function is approximately equal to (but always less than) the dangerous failure rate of the safety function. For a continuous mode safety function the hazard rate is equal to the dangerous failure rate of the safety function. Therefore, for both high demand mode and continuous mode safety functions, the hazard rate is a good measure of the safety performance of the safety function and so determines the SIL that in turn guides the selection of the measures used for the avoidance of systematic failures.

2.2.1 Hardware safety requirement

As for low demand mode safety functions, the reliability of the hardware has to be such that the random hardware failures do not cause the tolerable hazard rate to be exceeded. Again, this can be established by any of the recognised reliability modelling methods. However, at high demand rates (in relation to the proof test interval), the simplified equations used to estimate $PFD_{avg(achieved)}$ for low demand safety functions are no longer valid and give an over optimistic (i.e. low value) result. However, as shown below, the hazard rate resulting from random hardware failures in a safety-related system having an inherent reliability giving a dangerous failure rate, λ, cannot exceed λ. So the safety requirement will be satisfied if the reliability of the hardware is such that the overall dangerous failure rate is no greater than the tolerable hazard rate. This is the approach taking in the examples for high demand or continuous mode operation given in IEC 61508-6.

Alternatively, for demand mode safety functions, the hardware safety requirement can be expressed in terms of PFD_{avg}. This allows credit to be taken for periodic proof testing, thereby relaxing the hardware reliability requirement. However, to take advantage of this approach it is necessary to use more elaborate models to estimate $PFD_{avg(achieved)}$ than are used in the low demand region. Such models are described elsewhere (Sato, 1999).

2.2.1 Systematic safety requirement

In the case of high demand or continuous mode safety functions IEC 61508-1, table 3, gives the SIL which guides the selection of the measures adopted for avoidance of systematic failures.

2.3 Application of tables 2 and 3

Figure 3 illustrates how the definitions of 'low demand mode of operation' and 'high demand or continuous mode of operation' relate to the application of IEC 61508-1, tables 2 and 3. The definitions define the boundary between the two modes of operation in terms of the demand rate and proof test interval.

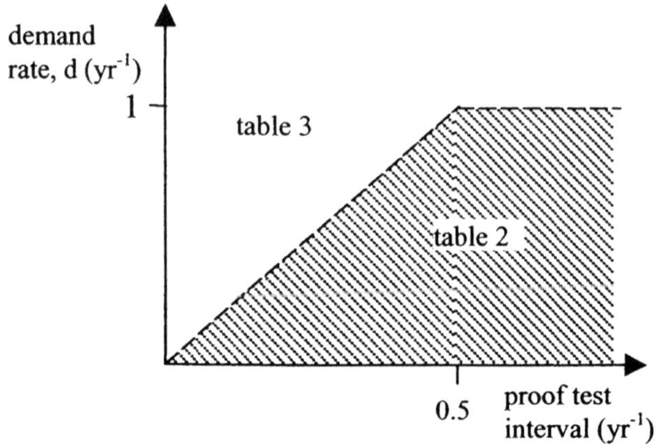

Figure 3 Application of IEC 61508-1, tables 2 & 3

3 System failure models

The rationale behind the definitions of 'low demand mode' and 'high demand or continuous mode' in IEC 61508 is based on the failure behaviour of a safety-related system due to random hardware faults. Underlying much of the reasoning is the distinction between safety-functions that only operate 'on demand' and those that operate 'continuously'. A safety function that operates on 'demand' has no influence until a demand arises, at which time the safety function acts to transfer the associated equipment into a safe state. A simple example of such a safety function is a high level trip on a liquid storage tank. The level of liquid in the tank is controlled in normal operation by a separate control system, but is monitored by the safety-related system. If a fault develops in the level control system that causes the level to exceed a pre-determined value, then the safety-related system closes the feed valve. With such a safety function, a hazardous event (in this case, overspill) will only occur if the safety function is in a failed state at the time a demand (resulting from a failure of the associated equipment or equipment control system) occurs. A failure of the safety function will not, of itself, lead to a hazardous event. This model is illustrated in Figure 4.

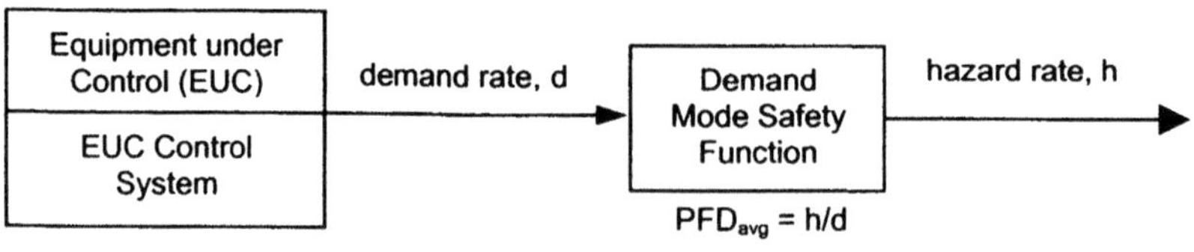

Figure 4 Demand mode safety function model

Assuming that the safety-related system in Figure 4 has a constant unrevealed dangerous failure rate due to random hardware faults, λ, then the mean time to failure (MTTF) of the protection system is:

$$MTTF = 1/\lambda \qquad (5)$$

If the safety-related system is subject to proof tests at interval, T, then, assuming that the product $\lambda.T$ is small, failures will occur, on average, half-way through the interval between proof tests. The probability of a demand occurring in the time T/2, P(demand in time T/2) is given by the Failure Probability Function,

$$P(\text{demand in time } T/2) = 1-e^{-d.T/2} \qquad (6)$$

But a hazard will only result if there is a failure of the safety-related system that is followed by a demand before the next proof test. The mean time to a hazard (MTTH) is therefore:

$$MTTH = MTTF/P(\text{demand}) \qquad (7)$$

And the hazard rate, h, is given by:

$$h = 1/MTTH \qquad (8)$$

$$= \lambda(1-e^{-d.T/2}) \qquad (9)$$

This is illustrated in Figure 5.

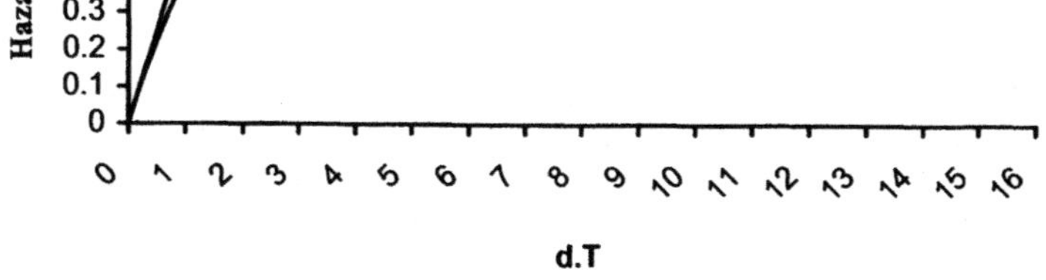

Figure 5 Hazard Rate for a demand mode safety function

As d.T increases, the hazard rate tends to the limit:

$$h = \lambda \qquad (10)$$

This is the high demand or continuous mode of operation. Here, λ is the measure of the safety performance of the safety-related system and is therefore used as the target failure measure which determines the SIL. This is the basis for the linkage between the hazard rate and SIL in IEC 61508-1, table 3 (see Figure 2).

At small values of d.T the hazard rate approximates to:

$$h \simeq \lambda.d.T/2 \qquad (11)$$

But for a safety function operating in demand mode:

$$h = PFD_{avg}.d \qquad (12)$$

This gives, for small d.T:

$$PFD_{avg} \simeq \lambda.T/2 \qquad (13)$$

This is the low demand mode of operation where the measure of the safety-performance of the safety-related system (which determines the SIL) is PFD_{avg}, approximated as $\lambda.T/2$. This is the basis for the linkage between PFD_{avg} and SIL in IEC 61508-1, table 2 (see Figure 1).

These 2 approximations intersect at the point:

$$d.T = 2 \qquad (14)$$

So for d.T<2, the approximation (11) gives the best result, whereas for d.T>2, the approximation (10) gives the best result. This is the rationale for the 'demand frequency no greater than twice the proof test frequency' boundary on the low demand mode of operation as currently defined in IEC 61508-4.

4 Demand rate boundary

The second boundary governing the application of tables 2 & 3 is at a demand rate of 1/yr. The rationale for this is explained by comparing the way in which the tables map to SIL for a demand mode safety function according to the demand rate and the required hazard rate. Figures 6 and 7 show tables 2 and 3 respectively in graphical form.

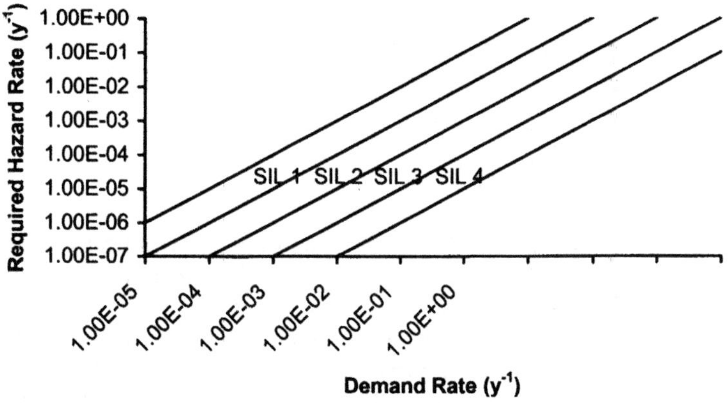

Figure 6 Graphical representation of IEC 61508-1, table 2

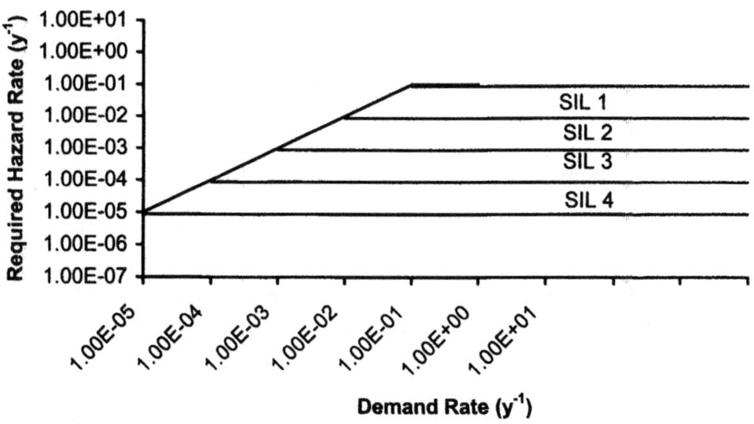

Figure 7 Graphical representation of IEC 61508-1, table 3

It can be seen that both tables give the same SIL, for a given hazard rate, at a demand rate of 1/yr. At demand rates lower than 1/yr, the SIL indicated by table 2 for a given combination of demand rate and hazard rate will, in general, be lower than the SIL indicated by table 3. Conversely, at demand rates higher than 1/yr, the SIL indicated by table 3 will, in general, be lower then the SIL indicated by table 2.

Therefore, in order to ensure a transition between the 2 tables that is free from a discontinuity in the indicated SIL, an additional boundary between the application of the tables is drawn at a demand rate of 1/yr.

5 Revision of IEC 61508

The IEC requires that all the standards that it publishes be periodically reviewed to identify the need for any changes in the light of the experience gained in the application of the standard or developments in technology. To this end, a consultation process through IEC national committees was started in 2001. This resulted in a significant number of comments relating to the definitions of modes of operation and the application of the SIL tables. A task team was set up as part of the overall revision process to consider these comments and produce recommendations for a future revision of the standard. Changes currently being considered include:

a) revising the definitions for 'low demand mode', 'high demand mode' and 'continuous mode' to emphasise the fundamental difference between demand mode and continuous mode safety functions. Demand mode safety functions have no influence on the EUC, or EUC control system, until a demand arises, whereas continuous mode safety functions retain the EUC in a safe state during normal operation;

b) removing the 'frequency of demands no greater than twice the proof test frequency' boundary between low demand and high demand modes. Whilst the ratio of demand rate to proof test frequency is a factor in the hardware reliability modelling for demand mode (estimation of $PFD_{avg(achieved)}$), there is no reason why it should be a factor in selecting which of the target failure measures (risk reduction factor or hazard rate) determines the SIL, and so it should not be a factor in deciding which of the tables should be used. In the case of demand mode safety functions, the risk reduction factor, and hence PFD_{avg}, is the measure of safety performance of the safety function and so it is the appropriate factor for determining the SIL for systematic measures, regardless of the ratio of demand rate to proof test frequency. However, the structure of the tables is unlikely to be changed so it will remain the case that, for a given demand rate and hazard rate, the SIL given by table 3 for a demand mode safety function will be lower than that given by table 2, at demand rates above 1/yr, and therefore the '1 demand/yr' boundary is likely to remain. This is consistent with the approach of IEC 61511, the process sector implementation of IEC 61508.

c) changing the description of the target failure measure for table 3, to 'hazard rate'.

6 Conclusions

In conclusion:

a) the rationale for the linkage between PFD_{avg} and SIL, as defined by IEC 61508-1 table 2, is that, for demand mode safety functions, the risk reduction factor required to achieve tolerable risk, and hence $PFD_{avg(target)}$ is the measure of the safety performance of the safety function. As the safety performance increases, the required SIL increases according to table 2, the aim being an increasing level of confidence that a systematic fault will not lead to a hazardous event.

b) the selection of the operating mode of a safety function in IEC 61508 should not be governed by the ratio of demand rate to proof test frequency as is the case according to the current edition of the standard;

c) the ratio of demand rate to proof test frequency is a factor in deciding the technique used for the hardware reliability modelling of the safety-related system. At low demand rates (in relation to proof test interval) it is permissible to use simplified equations based on the approximation $PFD_{avg} = \lambda.T/2$, whereas more elaborate models should be used at high demand rates;

d) the target failure measure in IEC 61508-1 table 3, is the limit of the hazard rate that achieves tolerable risk. This is not clear in the current edition of the standard;

e) demand mode safety functions are those which have no influence on the EUC or the EUC control system until a demand arises, whereas continuous mode safety functions are those which retain the EUC in a safe state during normal operation. The current definitions in IEC 61508 do not emphasise this important distinction;

f) the linkage between the required risk reduction or hazard rate required to achieve tolerable risk and SIL, as defined by IEC 61508-1 tables 2 and 3, is based on judgement as to what gives confidence that systematic failures, including those arising from specification errors, will not give rise to a hazardous event, taking into account the required risk reduction or hazard rate;

g) changes to IEC 61508 are currently being formulated as part of the IEC 61508 revision programme to clarify the standard in these areas. It is planned to publish a revised version of the standard, including these changes, in 2004 for consultation through the IEC national committees.

References

[IEC, 1998a] IEC 61508-1 "Functional safety of electrical/electronic/programmable electronic safety-related systems, Part 1: General requirements", 1998 (International Electrotechnical Commission)

[IEC, 1998b] IEC 61508-3 "Functional safety of electrical/electronic/programmable electronic safety-related systems, Part 3: Software requirements", 1998 (International Electrotechnical Commission)

[IEC, 1998c] IEC 61508-4 "Functional safety of electrical/electronic/programmable electronic safety-related systems, Part 4: Definitions and abbreviations", 1998 (International Electrotechnical Commission)

[IEC, 2000a] IEC 61508-2 "Functional safety of electrical/electronic/programmable electronic safety-related systems, Part 2: Requirements for electrical/electronic/programmable electronic safety-related systems ", 2000 (International Electrotechnical Commission)

[IEC, 2000b] IEC 61508-6 "Functional safety of electrical/electronic/programmable electronic safety-related systems, Part 6: Guidelines on the application of IEC 61508-2 and IEC 61508-3", 2000 (International Electrotechnical Commission)

[Sato, 1999] Sato Y, Misumi Y, "Estimation of average hazardous-event-frequency for allocation of safety integrity levels"" in Reliability Engineering and System Safety 66, 1999 (Elsevier Science)

Acknowledgement

The author would like to thank R I Wright, R M Consultants Ltd, for his contributions to this paper.

THE HUMAN SIDE OF RISK

Chasing Shadows: Science Journalism and the Politics of Risk

Stuart Allan
School of Cultural Studies
University of the West of England
Oldbury Court Road
Bristol, UK, BS16 2JP
E-mail: s.allan@uwe.ac.uk

The world of science, judging from some media portrayals, is a world of white-coated boffins peering through microscopes, laboratory benches with bubbling flasks set above flickering Bunsen burners, and racks of test-tubes and petri dishes emitting strange aromas. It is an insular world, cut off from the real world outside the laboratory window. The media tell us that this world of science is mind-numbingly boring and mundane in the repetition of its daily routines. Except, that is, for those rare moments when with a terrifyingly abrupt flash of insight (in the time it takes to, say, split an atom) the very future of humankind might suddenly appear to be hanging in the balance. If in the wrong hands science can be used to evil ends, in the right hands it is proclaimed to be our salvation. Scientists themselves tend to be represented as being decent, high-minded citizens tirelessly committed to the eternal pursuit of truth on behalf of their society. This endeavour, it follows, is a cornerstone of modern democracy, helping to make the world an organized, ordered system. Still, we are warned, there are exceptions. Lurking among their ranks are those intent on exploiting scientific knowledge for ominous purposes. These scientists, having been corrupted by greed or driven mad by a lust for power, are dangerously out of control.

The world of the media, at least according to statements sometimes made by scientists, is a superficial world driven by a frenzied obsession with entertainment over information, and with it style over substance. This is a world of smoke and mirrors, where nothing is as it seems, and where talk of ratings, target audiences and financial profits all but silences the voices of scientific truth. Journalists struggling to report on a scientific development, no matter how well intentioned they may be, will more often than not succumb to the forces of sensationalism to make their news account attract the public's wandering eye. If it bleeds, it leads. By the same logic, scientific facts must not be allowed to get in the way of a good story. Across the media, in-depth discussions and debates about scientific inquiries are up against, and losing out to, talking dinosaurs befriending humans, magicians happily breaking the laws of physics, mystics foretelling lottery results in their crystal balls, and horoscopes revealing people's fate and fortune as dictated by the stars. For some scientists, this recurrent misrepresentation of the scientific world by certain members of the media is more than just scandalous, it is contrary to the fundamental values and interests of democracy itself. The media's failure to give science the respect it deserves, they warn, will have dire consequences for the future.

It goes without saying, of course, that these two 'worlds' are being painted here with broad and colourful brushstrokes. My intention in doing so is to highlight how the often subtle boundaries demarcating what counts as 'science' in a modern society need to be situated in relation to the kinds of images one typically encounters in the media. Morc to the point, it seems to me vital that the contested limits of these boundaries be acknowledged from an array of different vantage points from across the science-media nexus. Precisely what science *is*, and what it is *not*, is anything but straightforward, as even a cursory glance at, say, a daily newspaper item about 'mad cow disease' or a television documentary about 'global warming' will immediately make apparent. Some might argue that science, like beauty, is very much in the eye of the beholder. In any case, to think of science and the media as separate worlds in constant collision with one another may be advantageous in some ways, not least with regard to identifying key sources of tension in public debates, but should not be understood too literally. Much more appropriate, in my view, is a critical engagement with scientific and media discourses that accounts for the complex ways in which they each strive to engender certain preferred ways of talking about the nature of reality. Such an approach recognizes that the extent to which their respective truth-claims converge or, just as importantly, are made to diverge from one another, will have a profound impact on our sense of the world around us (see also Allan 2002).

The line of inquiry I want to pursue in my paper today takes as its point of departure the thesis that efforts to discern what constitutes science must necessarily address the salience of media representations. To engage with several pressing issues which, in my view, go to the heart of efforts to better understand the changing imperatives of science journalism, I want to first turn to a recent Parliamentary report to help flesh out a critical line of inquiry.

Science and Society

'Society's relationship with science is in a critical phase,' declared a report by the Select Committee on Science and Technology, House of Lords, in March 2000. The report, titled simply *Science and Society*, drew upon evidence collected over a yearlong inquiry into the widespread perception that there is a serious crisis of public confidence in the biological and physical sciences and their respective technological applications. In the course of presenting its findings, the report identifies a range of issues which resonate, to varying degrees, in public debates across a range of different national contexts. As such, it will be used here as a springboard of sorts for the discussion to follow.

Underpinning this apparent crisis in public confidence is a paradox. At one level, the report states, 'there has never been a time when the issues involving science were more exciting, the public more interested, or the opportunities more apparent' (Select Committee on Science and Technology 2000). These claims are supported with reference to the results of recent opinion survey studies, as well as with regard to the growing salience of popular media output addressing scientific topics, among other factors. Nevertheless, on another level, 'public confidence in scientific advice to Government has been rocked by a series of events, culminating in the BSE [or 'mad cow disease'] fiasco; and many people are deeply uneasy about the huge opportunities presented by areas of science including biotechnology and information technology, which seem to be advancing far ahead of their awareness and assent.' Precisely how this enhanced level of public interest corresponds, in turn, with what the report describes as an 'increasing scepticism about the pronouncements of scientists on science-related policy issues of all types' is the subject of much debate.

In the view of the select committee's members, which included several distinguished scientists under the chairpersonship of Lord Jenkin of Roding, 'public unease, mistrust and occasional outright hostility are breeding a climate of deep anxiety among scientists themselves.' Interestingly, the report signals from the outset its commitment to exploring anew the varied sources of these tensions. Identified as being one of the more influential sources, as might be anticipated, is science journalism. Public attitudes to science, the report points out, are shaped by an array of institutions situated across the breadth of society, not least by the teaching of science in schools. Once people leave the education system, however, the news media become their principal sources of information about science. Therein lies the problem in the eyes of scientists. According to the select committee report, many scientists tend to be convinced that journalists 'have it in for them', that is, that the 'cherished freedom of the British press works against them.'

In order to unravel the complexities of science journalism, the committee distinguishes between three different types. First, there is the specialist scientific press, where news reports are written by scientists for other scientists. Second, there is the work of science journalists, namely specialist correspondents employed by mainstream news organisations. They will usually conduct their own journalistic research into a science story so as to ensure due accuracy in their handling of the facts. Third, there is the work of non-scientific correspondents. These journalists, mainly by dint of circumstance, typically find themselves writing about a scientific development as a general news story. In so doing, the report maintains, they may subject the story to 'a very different set of values and criteria' than might be otherwise expected from a specialist science reporter.

In comparing the activities of the latter two types of journalists, the report highlights several factors which together help to explain why the information about science being presented by the news media is not as effective as it might be. These factors begin with the simple observation that science reporters are first and foremost concerned with their role as a journalist, as opposed to that of an educator: 'Their primary aim, as with any journalist, is to get stories into the paper or programme, in fierce competition with other journalists.' This determination to see the scientific world in terms of news stories leads, in turn, to an undue emphasis being placed on those kinds of events that satisfy the news organisation's editorial policy.

Science and news, the report suggests, tend to be 'a poor fit'. This is so because 'newsrooms deal in simplified stories put together in haste, preferably with two opposing sides or views.' One witness before the committee is cited as recalling the following instructions of a BBC Radio 1 news producer prior to a live interview: '20 seconds, professor, and no long words.' Moreover, when priority is given to clashing viewpoints, especially when they are sensationalist ones, familiar notions of balance and fairness can be decisively undermined. The committee heard 'vehement criticism' of the practice, for example, where the news media 'give equal weight to the scientific consensus and to a minority view, whether in the interest of balance as they see it, or simply because confrontation makes good copy.' Particularly revealing is the report's contention that this crisis is at its most severe where scientific definitions of 'risk' are concerned.

'When science and society cross swords,' the report intones, 'it is often over the question of risk.' As it proceeds to point out, there are two dimensions to risk which are particularly significant in this context: 'the chance of something happening, and the seriousness of the consequences if it does.' How scientists choose to communicate their calculations of risk is not only a question of rigour and accuracy, but also one of politics. Acute difficulties may arise, for example, as soon as scientists undertake to formally quantify a risk, especially when they necessarily have to qualify their claims by acknowledging a degree of uncertainty ('It appears to be safe, on the basis of the following assumptions which require further research'). Confronted with such hedged assurances, journalists will more likely than not insist upon absolute certitude ('Is it 100 per cent safe?). Their inability to extract such an affirmation from scientists can sometimes lead them, in turn, to call into question the integrity of the actual risk calculation itself on these grounds alone. 'By this means,' the report suggests, 'the stage is set for confusion, cynicism and even panic.'

Not surprisingly, various individuals and groups seeking to help secure a place for these issues on the national agenda have warmly welcomed this report. Many publicly applauded its commitment to recasting several of the all too familiar premises underpinning debates about the science-media nexus. Finding particular support among those who regard the status quo as untenable was the committee's conviction that 'the culture of UK science needs a sea-change, in favour of open and positive communication with the media.' To date the government's response to the report has been positive, if largely limited to a formal reaffirmation of its importance as a contribution to ongoing debate. In particular, the report has been credited with influencing the development of policies set out in the Science and Innovation White Paper, published in July 2001. Disappointingly, if all too tellingly, the report's publication was virtually ignored by the very news organisations it set out to challenge. In the case of the broadsheet newspapers, for example, coverage typically consisted of a single item acknowledging the report's release, together with a brief sketch of its contents. It was then promptly consigned to the journalistic dustbin of history.

Re-defining Risk

My decision to dwell on the report's engagement with science journalism here is based on my sense that it succeeds – both by accident and by design – in effectively highlighting several pressing issues in need of further attention. Chief amongst these issues, as will be obvious from the discussion above, is the urgent need to rethink what we mean by 'risk' and how best to communicate it to members of the public.

'Risk', as we all know, has been the subject of considerable debate both within and beyond scientific communities, with meanings of the term often varying quite dramatically from one user to the next. In its technical sense, however, risk is usually defined as the calculated probability of an adverse consequence (such as a danger, harm or loss) arising because of a specific action or process. Adams (1999: 285) identifies three broad categories of risk:

- Directly perceptible risks: e.g. traffic to and from landfill sites;
- Risks perceptible with the help of science: e.g. cholera and toxins in landfill sites;
- Virtual risks – scientists don't know/cannot agree: e.g. BSE/CJD [bovine spongiform encephalopathy or 'mad cow disease' / Creutzfeldt-Jakob disease] and suspected carcinogens.

Self-described 'risk managers' tend to focus on the first two of these three categories. The reason, Adams (1999: 285) argues, is that '[q]uantified risk assessments require that the probabilities associated with particular events be known or be capable of plausible estimation'. Scientists, as many of them are quick to acknowledge, tend to frame issues of risk in terms of probabilities which are little more than confident expressions of uncertainty. 'When scientists cannot agree on the odds,' writes Adams (1999: 285), 'or the underlying causal mechanisms, of illness, injury or environmental harm, people are liberated to argue from belief and conviction' (see also Friedman *et al.* 1999; Scanlon *et al.* 1999).

Scientists' perceptions of risk, one study after the next suggests, can be at serious odds with those held by members of the public. Research commissioned on public perceptions of risk by Britain's Parliamentary Office of Science and Technology (1996), for example, provides a series of pertinent insights. In addition to the actual size of the risk, a variety of different factors are identified which appear to influence public perceptions:

- *Control* – People are more willing to accept risks they impose upon themselves, or they consider to be 'natural', than to have risks imposed upon them.

- *Dread and Scale of Impact* – Fear is greatest where the consequences of a risk are likely to be catastrophic rather than spread over time.

- *Familiarity* – People appear more willing to accept risks that are familiar rather than new risks.

- *Timing* – Risks seem to be more acceptable if the consequences are immediate or short-term, rather than if they are delayed – especially if they might affect future generations.

- *Social Amplification and Attenuation* – Concern can be increased because of media coverage or graphic depiction of events. Or reduced by economic hardship.

- *Trust* – a key factor is how far the public trusts regulators, policy makers, or industry. If these bodies are open and accountable – being honest, admitting mistakes and limitations and taking account of differing views without disregarding them as emotive or irrational – then the public is more likely to place credibility in them.

(Parliamentary Office of Science and Technology 1996)

These factors, taken together, contribute to a better understanding of why some risks are perceived as being more serious than others. Each in turn highlights, to varying degrees, the significance of the media in shaping these perceptions. That is to say, confronted by scientific uncertainty where risks are concerned, 'ordinary' or 'lay' members of the public are likely to turn to the news media, in particular, for a greater understanding of what is at stake. More often than not, news accounts will offer the assurance that a potential risk will remain uncertain only until further research and scientific investigation are able to provide the expected clarity and certitude (see also Adam 2000). Risk, not surprisingly, thus becomes a deeply political issue.

Explanations for this problem, in the opinion of some journalists at least, tend to revolve around the charge that most types of science fail the test of newsworthiness. Routine science, they tend to believe, is really rather boring. It lacks the stuff of drama necessary to spark lively newspaper headlines. At the same time, some scientists maintain that on those occasions when a certain scientific development is given due prominence, it all too frequently happens for the wrong reasons. Not surprisingly, they are quick to condemn instances of sensationalist reporting – where news considerations have given way to entertainment preoccupations – for misrepresenting the nature of scientific inquiry, and rightly so. 'Media wisdom has it,' writes Dunbar (1995: 147-148), 'that news must have impact and, especially, human interest to sell papers. But when science tries to compete with the social antics of the great and the not so good, it has only limited chances of success'. Not surprisingly then, as Nelkin (1995) observes, science typically appears in the press as 'an arcane and incomprehensible subject', around which there is a certain 'mystique' that implies it is to be properly regarded as a 'superior culture' with a 'distanced and lofty image'. Far from enhancing public understanding, she argues, 'such media images create a distance between scientists and the public that, paradoxically, obscures the importance of science and its critical effect on our daily lives' (1995: 15).

Accordingly, in examining the factors shaping how news organisations report on risk issues, a number of intriguing issues are brought to light. Science journalists will often claim that they simply follow their 'gut feelings', 'hunches' or 'instincts' when going about their daily work of identifying which risks (real or potential) are sufficiently 'newsworthy' to warrant public attention. Closer analysis quickly reveals that there are institutional imperatives which underpin these seemingly *ad hoc* judgements. To begin the work of unravelling these imperatives, then, it is necessary to recognise that scientific certainty is a discursive accomplishment. Of particular interest in this regard are the journalistic logistics involved in securing the co-operation of scientists, engineers and technicians to serve as news sources while recognising, at the same time, that they frequently have their own media agendas to pursue. The processes of selection indicative of one news organisation will be at variance with those of others, of course, but shared assumptions about these and related criteria of 'newsworthiness' recurrently underpin these daily negotiations.

Critiquing Science Journalism

Science journalists are charged with the duty of imposing meaning upon uncertainties, that is, it is expected that they will render intelligible the underlying significance of uncertainties for their audiences' everyday experiences of life in the 'risk society', to use Beck's (1992) evocative phrase. The process by which certain scientific developments are rendered 'newsworthy' while others, in contrast, are deemed unworthy of attention, is the outcome of a complex array of factors. Pertinent here is Hilgartner's (2000) examination of the strategies science advisors employ to achieve credibility for their proclaimed expertise – not only in the eyes of governmental officials, but also journalists. 'Science advice,' he writes, 'is a ubiquitous source of authority in contemporary Western societies' (2000: 4). More specifically, it is frequently mobilised so as to help reinforce the very legitimacy of state institutions, not least by separating public problems into separate 'scientific' and 'political' components. To the extent that this separation is sustained, Hilgartner argues, potentially controversial issues may be safely defused.

The complex process by which certain issues are politicised – or not – is likely to be contested from a number of different fronts. Needless to say, science advice does not automatically engender respect or approval, just as the authority of a given expert may be called into question by the next one. '[T]he claim that technical experts offer a value-laden vision has become a familiar idea,' Hilgartner observes, 'a cliché that stands in uneasy opposition to the even more commonplace notion that science-based expertise is universal and objective' (2000: 5). In his view, science advice is best conceptualised as a form of drama, so as to invite fresh questions about how it is produced, performed, and subjected to criticism. Expert advisors – together with their critics – each conduct their own individual performance, strategically presenting themselves by displaying a constructed persona in order to persuade people to accept the integrity of the claims they espouse. Each of these competing performers struggles to frame their chosen narrative on its preferred terms so as to be able to challenge, even pre-empt, the alternative narratives on offer. After all, almost any invocation of scientific rationality will spark the counter-charge of interest group politics being advanced by other means.

For scientists, engineers and technicians agreeing to serve as sources for a given news story, more likely than not they will promptly find themselves caught-up in the cacophony of claims and counter-claims. Of the myriad of concerns confronting them, several are particularly pertinent for our purposes here. Specifically, when interacting with a potential source, journalists will want to determine the answers to questions such as the following:

- Who is a credible, trustworthy and legitimate news source? Who is seen to possess sufficient expertise to interpret the possible threats, dangers or hazards associated with a risk in a reasonable manner? Whose scientific authority will anchor, in turn, the proclaimed impartiality of the news account?

- Wherein lies the news story? What will be of interest to my editor, fellow journalists, and audience? Where is the sense of conflict, drama and timeliness, all of which heighten the news value of the story? Does the story have a clear narrative structure, a straightforward beginning, middle and end, which will allow facts to be communicated quickly and easily?

- Is it safe? An unfair question to be sure, but one that will be asked as a matter of routine. Given that responses which fall short of 'yes' or 'no' will be treated as inherently suspicious, what sort of risk vocabulary can be drawn upon to communicate what's at stake in simple – but not simplistic – terms? How best to put a human face on scientific principles, let alone risk calculations?

- What happens when the experts disagree? Balanced reporting, by definition, means that there are always two sides to every story (the risk is either 'safe' or 'dangerous'). The greater the disagreement – and thus controversy – that can be engendered, the better. What is an acceptable risk, who decides, and by which criteria? If it eventually proves unacceptable, who will be blamed?

- What are the politics of risk? Who stands to benefit – and who will lose out – from the outcome of a particular risk decision? Who is going to be held accountable to resolve any ensuing risk crisis? How best to differentiate the public interest from what interests the public?

Expert sources, in contrast, may ask themselves questions such as:

- Why does this journalist want to talk to me? Where will I be made to fit in within the larger structure of the news item? That is to say, how will my views be aligned in relation to alternative views, how will the facts of the matter be contextualised? Can journalists be trusted to explain the story to the public in a manner that will avoid antagonising colleagues, let alone rivals?

- How to acknowledge that it is impossible to eliminate risk entirely without, at the same time, calling into question the authority of the risk assessment itself? Wherein lies responsibility for how risk assessments inform the decisions – political, economic, industrial and scientific – of others? Do scientists have particular moral obligations as citizens in a democracy? Who is entitled to criticise science, and who is not?

- In the absence of guarantees, how best to manage public fears of risk? How to judge which risks to avoid altogether, which ones may be worth taking, and how to explain the difference? In the course of identifying gradations of risk, how best to make them meaningful for the layperson? How best to make layperson perceptions of risk meaningful for scientists?

- Where risk calculations are concerned, what sort of evidence will be judged by the non-scientist to be compelling proof to sustain a claim, or sufficient grounds to challenge it? Who defines what counts as evidence, as well as the scientific consensus (where there is one) about its significance? Is it possible to separate out science from the attendant ethical implications of its use?

- How, and to what extent, are members of the public learning to live with the uncertainties associated with risk? Is risk an inevitable price to pay for scientific progress? Is risk aversion dangerous in itself?

In thinking about these and related questions against the backdrop of recent risk controversies – not least BSE and GM crops and food, but also Earth-bound asteroids, deep-vein thrombosis from air travel, brain cancer from mobile-phone radiation, the SARS crisis, and so forth – searching questions remain about how best to improve the reporting of risk and safety issues. Lessons need to be drawn from the mistakes made in the handling of these crises by journalists, of course, but also by those acting as the sources of their information and interpretation. A particularly pressing concern for all of those involved is the need to make the scientific language of risk and safety understandable to members of the public while, at the same time, facilitating the means by which their perceptions of risk inform the basis of the science involved. Ordinary or 'lay' people have a crucial part to play in scientific decisions, and their opinions have to be seen to count. 'Risks are not simply questions of abstract probabilities or theoretical reassurances,' as Richard Horton (1999), editor of the medical journal *The Lancet* points out. 'What matters is what people believe about these risks and why they hold those beliefs.'

It follows, then, that public trust is a prize to be won through open debate, and not something that can be assumed as a right by experts. In countries such as this one, as Harry Collins and Trevor Pinch (1998: 124) contend, 'the official response to public health risks has traditionally been paternalistic reassurance. The government judges that the danger of panic usually outweighs any real risk to its citizens'. Typically the main task becomes one of allaying public fears, a response that has clearly informed how the food crises have been handled. This tendency to pass the blame to science, emergent at a number of pivotal points as the respective BSE and GM food crises have unfolded, is arguably indicative of a cultural climate where science is frequently seen to be politics by another means. In this way, as Sheila Jasanoff (1998: 355) argues, these kinds of crises raise fundamental questions about the very legitimacy of state institutions and their advisory bodies. This is an age, she writes, 'of complex technological risks that defy full scientific understanding and paternalistic, topic-down control', hence the need to ensure adequate access for diverse voices in public dialogues about science. The policy process needs to be revitalised, she points out, so as to 'acknowledge multiple reservoirs of knowledge and expertise, recognize divergent perspectives on risk, and [to find ways to act] responsibly without necessarily knowing the full implication's of one's action.'

Democracy requires a robust exchange of viewpoints, and a journalism up to the challenge of giving them vigorous expression. New forms of dialogue about risk and safety issues need to be fostered, particularly where the absence of scientific certainty becomes controversial. It is vital that scientists, engineers and technicians be actively involved in the creation of consensus by experimentation and testing, but also with respect to sharing their assessments of the possible implications of a given risk for daily life. This is likely to mean that they will need to be far more proactive in the shaping of the very processes of science journalism. Knowledge of the imperatives underlying news production in general, and with respect to the reporting of risk and safety claims in particular, can help to close difficult gaps in mutual understanding. To further improve matters, responsibility for informing the public must be shared, with the corresponding lines of accountability clarified. Criticisms of science journalism ring hollow when some of those who might be otherwise enriching it with their insights – and, yes, their passions – refuse to play their part in public deliberation and debate.

References

Adam, B. (2000) The media timescapes of BSE news, in S. Allan, B. Adam and C. Carter (eds) *Environmental Risks and the Media*. London and New York: Routledge.

Adams, J. (1999) Cars, cholera, cows, and contaminated land: virtual risk and the management of uncertainty, in R. Bate (ed.) *What Risk? Science, Politics and Public Health*. Oxford: Butterworth Heinemann.

Allan, S. (2002) *Media, Risk and Science*. Buckingham and Philadelphia: Open University Press.

Beck, U. (1992) *Risk Society: Towards a New Modernity*. London: Sage.

Collins, H. and Pinch, T. (1998) *The Golem at Large*. Cambridge: Cambridge University Press.

Dunbar, R. (1995) *The Trouble with Science*. London: Faber and Faber.

Friedman, S.M., Dunwoody, S. and Rogers, C. L. (eds) (1999) *Communicating Uncertainty*. Mahwah, NJ: Lawrence Erlbaum Associates.

Hilgartner, S. (2000) *Science on Stage*. Stanford: Stanford University Press.

Horton, R. (1999) Genetically modified foods: 'absurd' concern or welcome dialogue? *Lancet*, 354: 1312.

Jasanoff, S. (1998) Book review, *Public Understanding of Science*, 7 (4): 354-356.

Nelkin, D. (1995) *Selling Science: How the Press Covers Science and Technology*, Second Edition. New York: W.H. Freeman.

Parliamentary Office of Science and Technology (1996) *Safety in Numbers?* Report 81. London: House of Commons.

Scanlon, E., Whitelegg, E. and S. Yates (eds) (1999) *Communicating Science, Reader 2*. London: Routledge.

Select Committee on Science and Technology, House of Lords (2000) *Science and Society*. London: HMSO.

Completing the Human Safety Jigsaw

Richard G. Scaife
The Keil Centre Limited,
Chartered Psychologists
Edinburgh, Scotland

Abstract

This paper describes an approach to the integrated assessment of human error, violations and safety culture which is intended to overcome some of the gaps which can exist when separate aspects of human factors are analysed in isolation.

1 Introduction

In any safety-critical industry the investigation of incidents, accidents and near-misses is of paramount importance. The means to perform such investigations have developed over a number of years, and most have their roots in the analysis and investigation of failures of hardware or software systems. More recently, however, the reliability of engineered systems has been seen to improve, whereas the reliability of the other key part of any system – the human – has remained relatively unchanged. The net effect of this trend has been that failures of human performance have been elevated in importance in many industries, which in turn has prompted the development of techniques to improve the level of rigour applied to the investigation of human performance and behaviour.

There are a wealth of tools, techniques and methodologies available that allow the analysis and investigation of the whole range of human behaviour and performance in relation to accidents, incidents and near-misses. They cover organisational influences on behaviour such as safety culture and individual factors such as human error and violations. Up until fairly recently, such tools, techniques and methodologies were designed by specialists for specialists, and were often difficult, if not impossible to use without lengthy formal training in Psychology or Human Factors. However, fuelled by the change in balance between technological and human failures, such tools have been developed targeted at the incident investigator. A key benefit of this development is that investigators can capture data on human performance and behaviour much earlier in the analysis process.

The tools for the analysis of human error, violations and safety culture have been introduced within industry either during an investigation or at some point following the investigation. However, generally speaking such tools are used in isolation, rather than as an integrated tool-kit to investigate all aspects of human behaviour and performance.

In most industries the investigation of an incident, accident or near-miss starts as soon as possible after the event (within minutes or hours) to ensure that the evidence of what has happened is still fresh. When human factors analysis is conducted this often occurs after the main evidence collection phase of the investigation, and so may miss some of the more volatile evidence that is only available for a short time after the event.

This paper describes the development of a tool set for incident investigation that incorporates the use of a suite of tools to assess human error, violations and safety culture as an integral part of the investigation. This approach helps to overcome issues associated with a delay in the investigation of human factors that can occur if such techniques are applied in a stand-alone context. This approach also has the benefit of collecting data on the human aspects of safety as a whole, rather than independently conducting several forms of analysis. This allows the complex relationships between people, the organisation, the environment and the task to be captured.

Table 1 summarises the coverage of human factors afforded by the integration of safety culture, violations and human error analysis methodologies.

Type of Analysis	**Safety Culture**	**Violations**	**Human Error**
Timescale to Fix	Long Term	Short to Medium Term	
Level of Analysis	Organisational and social influences of behaviour.	Work group influences on behaviour	Task and cognitive influences of behaviour
	ORGANISATIONAL	**GROUP / TEAM**	**INDIVIDUAL**

Table 1: Summary of Analysis Techniques and Coverage of Human Factors

2 The Industry Requirement

A number of organisations regularly conduct safety culture assessments to identify issues at the organisational level that impact on safety performance, and to determine how to overcome any problems that are brought to light. Some of these organisations also analyse human behaviours to determine why individuals or work groups violate procedures, and how these people could be influenced to adopt new behaviours. A sub-set of these organisations also use human error analysis to determine why errors occur and how their impact could be reduced.

In some organisations all three approaches are applied, but seldom in the context of incident investigation, and seldom are they all considered together to build a comprehensive picture of human safety.

There is sufficient evidence from well-known historical events to indicate that safety culture (Piper Alpha), human error (Kegworth) and violations (Herald of Free Enterprise) all play major roles in the occurrence of incidents. Coupled with the increasing importance being afforded to the human factors causes of incidents, organisations have a very convincing case for the development of an integrated suite of tools for incident investigators.

Many organisations already use root cause analysis techniques that allow the identification of critical factors in the occurrence of incidents. For each critical factor, related behaviours can be isolated. A number of these factors can be related closely to human error, violation or safety culture. Formal approaches are required to perform in-depth analysis of such factors and determine the root of human factors problems and how they could be dealt with.

In order to meet the objective of understanding the root causes of human behaviour and performance, it is clearly desirable to provide investigators with an integrated tool set that will do just that. Any such approach has to be made up of proven methods and techniques that can be integrated in some meaningful way. The integrated approach must be able to tell the investigator why the individual(s) involved in the incident behaved in the way that they did, and what could be done to either prevent them and others from doing so in future, or to reduce the probability that they will do so, or to reduce the impact in the event that the same circumstances recur.

The following sections describe an approach developed by The Keil Centre based upon three existing human factors analysis tools. The tools are described first, followed by a description of the approach taken to integrate them for use by incident investigators.

3 The Tools

3.1 Safety Culture

A vast array of methods have been developed for the assessment of safety culture, with varying degrees of success. One of the reasons why some of these methods have been unsuccessful was their failure to include the employees in the development of action plans to improve performance. They also tend not to consider the maturity of safety culture where the assessment method is deployed, and therefore can result in actions being formulated that may be inappropriate for that particular site.

In 1999 A joint Health and Safety Executive (HSE) and oil industry-funded project to address some of these concerns led to the development of a Safety Culture Maturity® Model[1] (SCMM) (see for example Lardner, Fleming and Joyner (2001). The SCMM is based on the capability maturity model concept, initially developed by the Software Engineering Institute of Carnegie-Mellon University (Paulk et al, 1993), as a mechanism to improve the way software is built and maintained. The SCMM aims to assist organisations in (a) establishing their current level of safety culture maturity and (b) identifying the actions required to improve their safety culture.

The components of the SCMM were based on the safety culture features listed in the Health and Safety Executive's human factors guidance document HS(G)48 (HSE, 1999). The initial model was tested by interviewing safety experts, operational managers, safety representatives and frontline staff about their company's safety culture development and the applicability of the SCMM. This led to the definition of a Safety Culture Maturity® Model, with five levels of maturity (as shown in Figure 1) and ten elements, namely:

- Visible management commitment
- Safety communication
- Productivity versus safety
- Learning organisation
- Health and safety resources
- Participation in safety
- Risk-taking behaviour
- Trust between management and front-line staff
- Industrial relations and job satisfaction
- Safety training

The SCMM presented in Figure 1 below is set out in a number of iterative stages. It is proposed that organisations progress sequentially through the five levels of maturity, by building on their strengths and removing the weaknesses of the previous level.

The Keil Centre has developed the Safety Culture Maturity® concept into an assessment tool to measure ten key elements of Safety Culture Maturity®. An interactive workshop process allows employees to identify current levels of Safety Culture Maturity® and what needs to happen to move to the next level.

The Safety Culture Maturity® model is a participatative, solution-focused safety culture assessment and improvement method. What makes it different from other safety culture assessment processes is that it has a strong focus on solutions, involves a high degree of workforce participation, and provides an opportunity for

[1] Safety Culture Maturity® is a registered trademark of The Keil Centre Limited

staff at all levels to learn more about key elements of safety culture, and their role in its development and maturity.

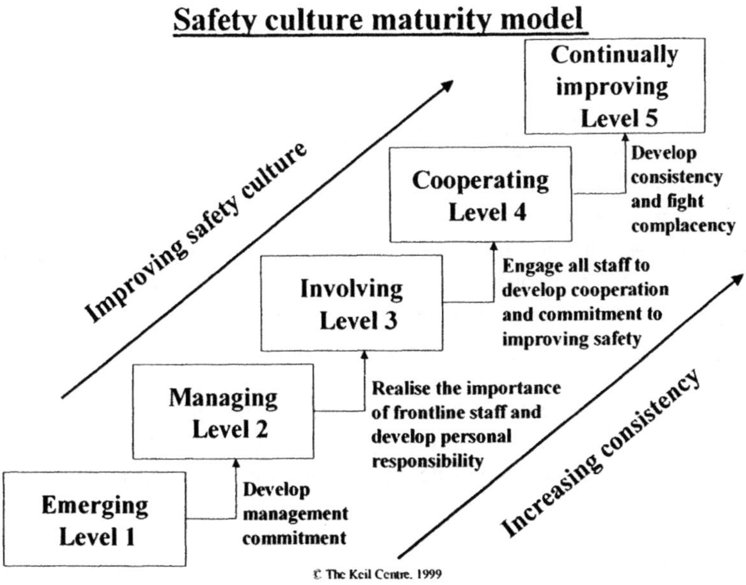

Figure 1 : Safety Culture Maturity® Model

3.2 Violations

ABC analysis (e.g. Daniels, 1999, Fleming and Lardner, 2002) is a tried and tested technique for understanding why people intentionally behaved as they did (in this case violate a working practice or procedure). The aim is to identify how this behaviour can be exchanged for another desired behaviour. It is applicable to any intentional behaviour, not just safety behaviours.

The ABC model is so called because of the three elements involved in understanding why people intentionally behaved as they did:

A - refers to **A**ntecedents, which come before the behaviour and prompt or trigger behaviour
B - refers to the specific **B**ehaviour we are interested in
C - refers to the **C**onsequences of that behaviour for the person involved

The ABC model assumes the following 3 propositions are true-

- Behaviour is largely a function of its consequences
- People do what they do because of what happens to them when they do it
- What people do (or do not do) during the working day is what is being reinforced

Most unsafe behaviours do not involve people deliberately intending to harm themselves or others. From their point of view, their behaviour made perfect sense. ABC analysis helps the investigator understand, from the other person's point of view, the antecedents (which triggered the unsafe behaviour), and consequences (which reinforced the unsafe behaviour). Once this is understood, antecedents and consequences can be rearranged (and written into recommendations) in such a way that will make it more likely that the person involved and others will behave safely in the future.

Antecedents' role is to get the behaviour to occur in the first place. Consequences' role is to get the behaviour to continue to occur. Much of traditional Health, Safety and Environment (HSE) activity is devoted to providing antecedents for acceptable behaviours.

Intentional safety violations can be grouped into 3 categories:

Routine violations	-	where breaking a rule or procedure has become a normal way of working within the work group
Situational violations	-	where breaking a rule is due to pressures from the job such as time pressure, low manning, high workload, inappropriate equipment, weather
Exceptional violations	-	When things go wrong, breaking a rule to solve a problem, even though aware that a risk is being taken

An ABC analysis begins by defining the antecedents of the behaviour. Antecedents can be the presence or absence of factors such as suitable tools and equipment, other peoples' example and procedures.

After the antecedents have been defined, the consequences of the behaviour are described from the perspective of the person who was involved. Examples of consequences include getting injured or harmed, saving time and getting approval from a supervisor or manager.

Each consequence is then assessed for the following, from the perspective of the person who performed the behaviour:

Positive (P) / Negative (N)	– from their perspective, if this consequence occurred, would it be positive or negative? Note that getting injured or harmed will usually be assessed as negative.
Immediate (I) / Future (F)	– from their perspective, does this consequence occur immediately after the behaviour (now or soon) or in the future? Note that getting injured or harmed will usually be assessed as something that will happen in the future, not today.
Certain (C) / Uncertain (U)	– from their perspective, is it relatively certain that this consequence will occur, or somewhat uncertain? Note that getting injured or harmed will usually be assessed as something which is uncertain (i.e. it has not happened to me yet, so it won't happen today).

Positive, Immediate and Certain consequences influence behaviour much more strongly than Negative, Future and Uncertain consequences do.

Having fully described the problematic behaviour, the next step in the process is to define a safe alternative to this behaviour, which antecedents will help to ensure that this behaviour happens, and the consequences that will help to reinforce the behaviour. The results of the analysis can then be turned into practical recommendations to reduce unsafe behaviours and introduce new alternatives to replace them.

3.3 Human Error Analysis

Various forms of human error analysis have been widely used in a number of industry sectors, although the specific tools used for the purpose vary. The tool that is described here is based on an approach developed for use in Air Traffic Control, and was designed to integrate into the existing incident investigation process (Shorrock, 2003). Unlike many tools, this form of HEA was designed to be used by incident investigators, not human factors specialists and so assumes little or no knowledge in the field of psychology and human behaviour.

The model used assumes that an error will be the result of a failure in one of four areas of human cognition and performance, as shown in Figure 2. The human operator perceives information about the outside world using all of the senses, uses this information along with information retrieved from memory to arrive at decisions that are used to determine the actions that will be taken.

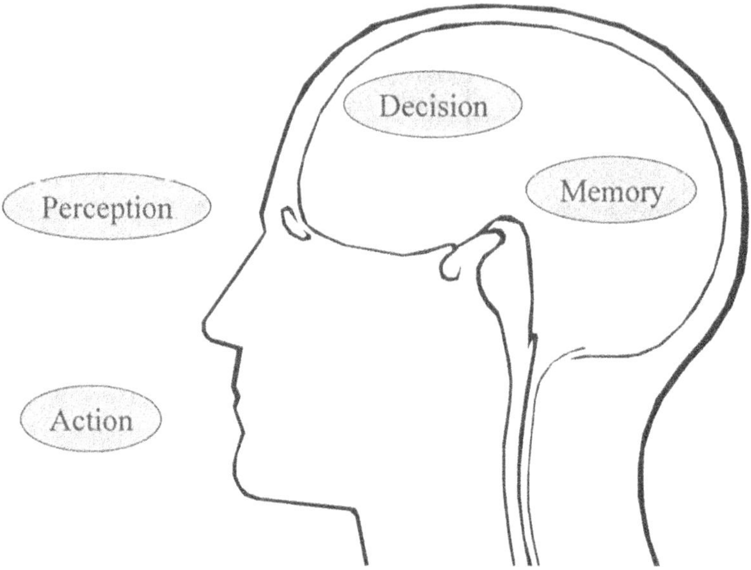

Figure 2 : Model of Human Cognition (After Wickens, 1992)

To find out why these types of error happen, it is necessary to work out what caused the failure in that part of the system, i.e. what were the underlying psychological factors? As well as telling us why an error has occurred, the underlying psychological factors also give us strong clues as to what we can do to reduce the impact of errors of this type.

It is also necessary to be mindful of the fact that human performance in general is very heavily influenced by the conditions under which the operator performs. These conditions are known as performance shaping factors, and can help to further clarify why an error occurred, and also provide a great deal of extra information to help specify a practical solution.

4 Integrating the Tools

Whilst the tools described above undoubtedly have their own benefits when used in isolation, these benefits can be greatly increased by integrating all three methods within an existing incident investigation methodology. The analysis of errors and violations can be used during an investigation to identify short to medium term actions to improve safety. The assessment of safety culture can be used to identify longer term improvement actions with wider benefits, and is therefore likely to be used following an investigation, rather than as an integral part of the investigation itself. However, collection of data on safety culture issues should take place during the investigation to ensure that the information collected is fresh from the event.

To effectively integrate these approaches, some form of overarching process is required to provide a framework which will guide the user to conduct the most appropriate form of analysis. One option for such a structure is Root Cause Analysis (RCA). The process of conducting RCA is such that the incident is investigated from the top down, progressively revealing more details of the causes.

It is therefore possible, once the causes of the incident have been defined, to provide guidance to the investigator to help determine whether a behaviour represented an error or a violation, and whether there may be issues relating to the safety culture of the organisation. A preferred safety culture model can be mapped onto the RCA model.

Practical guidance is required to classify a behaviour as intentional (a violation) or unintentional (an error). Having been through the RCA process, an investigator should have sufficient evidence from interviews, witness statements, and so on to make this judgement. The integrated tool set also needs to be flexible, so as to allow the investigator to change from analysis of an error to analysis of a violation, and vice versa. It is possible to begin an analysis believing that a behaviour was intentional, and therefore a violation, only for it to transpire that the behaviour was the result of an erroneous decision. It is also very possible that an incident involves both an error and a violation as part of the chain of events, hence providing sufficient guidance to help the investigator distinguish between them is very important.

A question which arises during any investigation is whether the particular circumstances of the incident in question are symptomatic of wider failings in the site or organisation's safety culture. The investigation team does not typically have the time to answer this question. It is also possible that a series of incidents, when taken together, can indicate wider failings in the site or organisation's safety culture. Furthermore, investigators may be reluctant to make recommendations about wider aspects of safety culture.

To determine whether it is appropriate to conduct an assessment of safety culture, it is necessary for the investigator to be able to determine the signs that there may be such a problem. It is possible to cross-reference the elements of safety culture to causes in a standard RCA approach, and to turn these cross-references into a checklist for use by the investigator. This provides a direct link for the investigator from the analysis technique with which they are working to safety culture assessment, indicating the areas of safety culture that are likely to require most attention.

The data for this activity are captured during the incident investigation, however it is extremely unlikely that any form of safety culture assessment would take place whilst the investigation were still active. It is anticipated that if the checklist revealed that there were a large number of causes linked to the Safety Culture Maturity® Model then there would be a recommendation arising from the investigation that safety culture be assessed at some future date. It is also possible that the data from a number of investigations would be examined for trends, and if any such trends were present, then a safety culture assessment may be recommended.

Similar approaches to link RCA to ABC analysis and HEA can be taken.

In the case of ABC analysis, for example, deficiencies in work tools and equipment could act as an antecedent for unsafe behaviour. The same can be done to identify nodes of an RCA technique that could relate to the consequences of problematic behaviours. For example, a RCA technique may include a potential cause relating to occupational stress, and the reduction of stress may be a consequence of problematic behaviour that helps to reinforce that behaviour.

All nodes of a RCA methodology can be examined to identify those that could relate to the antecedents and consequences of problematic behaviours. This information can be used in the form of checklists by incident investigators to lead them into the ABC analysis from their initial investigation using RCA.

In the case of HEA, causes such as poor judgement which are commonly included in RCA techniques are indicative that there may be a human error component in the initiation or escalation of the incident. In the same way, these can be identified and listed to provide the investigator with a route map to the conduct of human error analysis. Key parts of any human error analysis are the performance shaping factors that can influence the occurrence of errors. In most forms of RCA, such factors are well represented in the form of environmental conditions, equipment conditions, and so on. Many of these reflect categories of Performance Shaping Factor (PSF) contained within HEA techniques such as TRACEr (Shorrock, 2003). It is therefore possible to provide guidance by mapping PSFs to factors in the RCA technique.

The use of RCA to draw together these human factors analysis tools in this way requires some form of procedure to be written around the associations between the RCA technique itself and the tools. Such a procedure needs to provide clear steps for the identification of the behaviours involved in the incident, and guidance on the classification of those behaviours as intentional or unintentional. This leads the analyst into the appropriate form of analysis, leading to the development of appropriate corrective actions.

The procedure also needs to provide guidance on the identification of potential safety culture issues for follow-up after the investigation. The procedure should not be aiming to provide guidance on how to conduct the safety culture assessment, just to identify the issues that require attention.

Since this integrated approach is intended for use by incident investigators who may not have received any formal human factors training, it is useful to provide guidance on the generation of actions, or recommendations for action from the analysis. Such guidance might include:

- How to specify and introduce a new behaviour to replace an existing unsafe one based upon the use of ABC analysis;
- Example solutions to common types of human error identified using HEA and how to introduce them;
- Example interventions that can be used to improve levels of trust, participation and other key factors that contribute to a good safety culture.

5 Summary

The human factors issues surrounding incidents can be complex: they are seldom purely down to a violation of a procedure, or an error, or a problem with the way that safety is handled within the company. More often it is a combination of all three. Traditional investigation techniques help to identify what happened in terms of human factors, but they do not tend to take the analyst deeper to identify the reasons why they happened. Consider a jigsaw puzzle of a seaside scene with golden sandy beach, blue-green sea and bright blue sky, representing the human factors issues of human error, violations and safety culture. Conducting a conventional safety analysis will be akin to putting all of the edge pieces into the puzzle. This will give you an indication of where the sky, sea and sand lie within the picture, but little more.

Given the relative reliability of engineered systems and human operators, piecing together why humans violate procedures, commit errors and how they are influenced by safety culture is vitally important. The development of specialist tools for just these reasons helps in the investigation of safety culture, human error and violations, but they tend to be used independently, and by specialists after the event rather than being integrated and used by investigators during the investigation itself. This is akin to completing the sky, sea or sand parts of the puzzle, giving a good picture of one part of the puzzle, but leaving the other parts incomplete.

This paper focuses on the use of Human Factors analysis techniques in the investigation of incidents, and therefore concentrates on the retrospective application of these techniques. It is just as important to put in place prospective measures to manage safety culture, violations and human performance and to reduce the impact of such factors in future incidents. Typically such an approach would involve:

- A behavioural safety programme to continually reduce the potential for violations by discouraging unsafe behaviours and encouraging safe behaviours;
- Predictive error analysis as part of the procurement and development lifecycles to identify potential influences on error and reduce or remove them;
- Periodic safety culture assessment to maintain identify potential problems and formulate solutions.

The ultimate goal is to be able to integrate all three types of analysis tool into an incident investigation process, allowing the consideration of all human factors issues together in developing the overall picture. This is akin to being able to complete all of the parts of the puzzle, and to be able to see the interactions between the three aspects of human factors.

Such an approach has been developed by The Keil Centre, and is about to undergo a period of testing and evaluation. It is hoped that the results of this testing and evaluation would form the basis of a future paper.

6 References

Daniels, A. (1999). Bringing out the best in people. McGrawHill. New York

Fleming, M. and Lardner, R (2002). Strategies to Promote Safe Behaviour as Part of a Health and Safety Management System. CRR 430/2002, HSE Books, Norwich UK.

HSE (1999) HS(G)48 Reducing error and influencing behaviour, Norwich HMSO

Lardner, R., Fleming, M. and Joyner, P (2001); Towards a Mature Safety Culture. In Hazards XVI, Analysing the Past, Planning the Future. Institute of Chemical Engineers, Rugby UK.

Paulk, M.C., Curtis, B., Chrissis, M.B. & Weber, C.V. (1993) Capability Maturity Model, Version 1.1. *IEEE Software* 10 (4) 18-27.

Shorrock, S.T. (2003). The Retrospective and Predictive Analysis of Human Error in Air Traffic Control. Doctoral Thesis, University of Nottingham.

Wickens, C.D. (1992). *Engineering Psychology and Human Performance.* 2nd Edition. New York: Harper Collins.

ASSESSMENT AND THE DERIVATION OF EVIDENCE

Evidence Gathering Using Static Code Analysis

Chris Sampson
Advantage Technical Consulting
'The Barbican', East Street, Farnham, Surrey, GU9 7TB, UK

Abstract

Static Code Analysis (SCA) has a proven track record as a powerful software verification technique providing the necessary rigour for safety-related software. A number of mature tools supporting SCA are available. However, static analysis also has a reputation as being costly and labour-intensive. This paper looks at recent advances in identifying objectives and processes for SCA and assesses the potential for such analyses to provide, in conjunction with new software safety standards such as the CAA's SW01 and Ministry of Defence's DEF STAN 00-56 Issue 3, a cost-effective and focussed method of gathering evidence that software performs safely.

1 Introduction

The use of software in systems brings many advantages and equally many risks. An important risk, perhaps the most important, is that a software failure may lead to injury, illness or loss of human life. Where this risk is credible, software is known as safety-related software.[1]

The dependability of safety-related software is therefore an important concern. Developers of safety-related software and organisations procuring such software should be aware of the issues involved with the safety of their software, should recognise that software safety cannot be devolved from system safety, and should manage the safety of the system and its software. (UK MoD, 2003) identifies the following major stages in a safety management programme:

1. Hazard Identification,
2. Risk Analysis,

[1] The term "safety-critical" is also widely used, typically when loss of life can occur.

3. Risk Assessment,
4. Risk Reduction,
5. Provision of evidence.

Although this process, which is well established for system safety management, can be used for management of safety-related software, it tends not to be. Instead software assurance, following guidance given in the standards, has traditionally followed a process-based approach. Process-based assurance, however, has several serious drawbacks. Firstly, the prescribed processes only weakly integrate with system safety management approaches. Secondly, they can be expensive to apply, with the consequence that frequently only a subset of the process is applied. Thirdly, the combination of SILs and processes encourage an unjustified reduction of rigour in design and verification techniques for lower SILs. Fourthly, process based standards are not well suited to the use of commercial off-the-shelf software (COTS). Support for such process-based standards is seeping away, as evidence-based approaches - such as that listed above - gain support.

Static Code Analysis (SCA) is one of the basket of software assurance techniques recommended by process-based standards such as DEF STAN 00-55 Issue 2 and IEC 61508. As we see in Section 2 , SCA is a collective term for a variety of techniques, with differing objectives and degrees of rigour. Although the historical use of SCA has been somewhat divorced from the early stages of the safety management process, this paper argues that SCA, particularly in its most rigorous forms, is well suited to the evidential approach. Indeed, the evidential approach offers the potential for more focussed SCA, reducing the high cost traditionally associated with SCA.

2 Static Code Analysis projects

A review of the literature reveals a number of varying and contradictory understandings of static code analysis. A number of texts describe checking control flow and data flow properties such as unreachable code and uninitialised variables. Others concentrate on the measurement of software metrics such as McCabe's metric. Yet more papers describe detailed formal analysis of properties of the code. Often, techniques are defined in terms of the capabilities of software tools. Furthermore, there seems to be some confusion about the objectives of SCA. The following definition is from Wikipedia on-line encyclopaedia.

> "a set of methods for analysing software source code in an effort to gain understanding and to target areas for review and/or rewrite"

We certainly would hope to avoid the need for further review after SCA.

The market place is crowded with a plethora of tools claiming to perform "Static Code Analysis". Language-specific code checkers such as PC-Lint can be called static analysis tools. Dynamic Test Tools such as LDRA Testbed and Adatest have "static analysis" elements. Language analysis tools such as Polyspace and Merle perform checks, with a high degree of automation, that the Weakest Integrity Preconditions of language constructs are satisfied by the program. Finally, tools such as Malpas and Spark enable user-defined assertions to be written in a tradition pre- and post-condition form and proved.

Two themes are common to most of the understandings and tools:

- it is applied at source code level,
- it analyses the source code without executing the code

Clearly a better understanding of what SCA can offer is required before the technique can be recommended for use on safety-related software.

A review of a sample of safety-related software projects on which SCA has been used is helpful in coming to a better understanding of the technique.

2.1 Darlington Reactor Shutdown software

The Nuclear Power Plan (NPP) at Darlington was the first Canadian NPP to use a software based reactor shutdown system. The software was assessed in the late 1980s by Ontario Hydro using a manual static code analysis technique developed by Dr David Parnas (Archinoff, 1990). The approach uses "Function-Program" tables to derive code behaviour from the source code and compare against the software design specifications. The SCA was part of a wider verification process that included traditional unit, integration and system testing as well as a software hazard analysis. The SCA activity was independent of the software hazard analysis.

The Darlington SCA project was extremely labour intensive, analysing 26000 lines of code in 35 man years. It was realised by the Darlington team that tool support was necessary in order "to be practical". However, part of the benefit of the Darlington analysis was seen to be derived from the human involvement in the analysis. A totally automated anaysis process could be counter-productive.

2.2 Sizewell 'B' Primary Protection System

The Primary Protection System (PPS) for the Sizewell 'B' Nuclear Power Plant (NPP) was the first safety-critical software based system to be used in a UK NPP. The software was developed by Westinghouse Electric and consisted of approximately 100,000 lines of software written in PL/M-86 and ASM-86 (a structured version of 8086 assembler). (There were additionally a further 100,000 lines of configuration data.) The software, not including the configuration data, was subjected to retrospective Static Code Analysis (i.e. the software was written before the analysis was performed, and was not designed with ease of analysis in mind) using the MALPAS toolset (Ward, 1993). The analysis approach was extremely rigorous, consisting of Control Flow, Data Use, Information Flow, Semantic and Compliance Analysis. The Compliance Analysis stage was the most rigorous and provided a formal verification, using Pre- and Post-conditions, that the source code conformed to software design specifications. In order to achieve this, the source code had to be translated into Intermediate Language, the input language to MALPAS, and the English language specifications were converted into the Function and Term-rewriting rules used by the MALPAS tool.

As with the Darlington project, the SCA activity was part of a wider V&V programme of work consisting of Unit, Integration and System testing as well as code walkthroughs. A high-level hazard analysis of the PPS design was used to remove non safety-critical subsystems from the scope of SCA.

The Sizewell SCA project was also labour intensive, analysing 100,000 lines of code in 80 man-years between 1989 and 1993. Productivity was hindered by the slow processing power available at the time, analysis was performed on VAXes - Compliance Analysis of a single software procedure could take up to 12 hours; by the difficulty in abstracting information, the analysis process did not use information hiding; and in the difficulty in modelling in IL complex data structures, complexity that was hidden in the source code by the use of pointers. However, the Sizewell analysis did prove the feasibility of using SCA on large embedded systems, and was instrumental in the successful licensing of the system by the Nuclear Installations Inspectorate.

2.3 C-130J Hercules Avionics software

In 1994, the UK MoD made a decision to purchase a new variant of the C-130 Hercules transport aircraft (Harrison, 1999). The new version of the aircraft contained a new software-based avionics system, which was divided into a number of Line Replaceable Units (LRUs). RTCA/DO178-B was adopted as the standard to follow; however, the requirements of the standard were extended to include SCA of all software classified as Level A and B[2]. 23 LRUs were deemed to require SCA, totalling 500,000 lines of code. The software was written in a number of languages, including Ada (one-third of the Ada software was Spark-Ada compliant), C, LUCOL, PL/M and various Assemblers.

MALPAS and the Spark Examiner were used to perform SCA. A four-stage Goal Directed approach was developed:

---

[2] RTCA/DO178-B classifies safety-related software into 5 Levels, named A to E, with Level A being the highest level of safety.

- Goal Identification
- Static Code Analysis
    - Software analysed with MALPAS was subjected to Translation, Control Flow, Data Use, Information Flow and Semantic Analysis. Compliance Analysis was not performed.
    - Software analysed with the Spark Examiner was subjected to Control Flow, Data Use, Information Flow, generation of Verification Conditions and Run-Time check information. Verification Conditions were discharged using the algebraic simplifier, providing a check that the code meets its specification. The specifications were written in the semi-formal CORE language (which is a similar approach to the Program/Function tables used on the Darlington project). The use of a semi-formal specification style eased the code/specification comparison.
- Sentencing (i.e. categorisation of anomalies). The results of the "Goal-Directed" stage were used to filter anomalies that would not have an impact on system safety.
- Regression Analysis.

The project is reported as having analysed 500,000 lines of code in 70 man years (Nadjm-Tehrani, 2002).

2.4 Ship/Helicopter Operational Limits Instrumentation (SHOLIS)

SHOLIS (Chapman, 2000) is a software-based system that advises ship's crew on the safety of helicopter operations under various scenarios. The software was developed in accordance with DEF STAN 00-55 (Issue 2). A software hazard analysis was performed and on this basis certain parts of the software were designated as safety-critical. Safety critical software was formally specified using Z, developed in Spark Ada, and a "partial correctness" performed of the code against the specification. Information Flow analysis was used to demonstrate functional separation of critical and non-critical software. Freedom from run-time exceptions was demonstrated for all code. Static analysis of I/O usage, memory and timing was used to show separation of non-functional properties. Finally, proof that the system's top-level safety properties were maintained by the software was carried out.

2.5 European Aeronautics Defence and Space (EADS) Launch Vehicle

The software for the EADS Satellite Launch Vehicle contains 100,000 lines of mission-critical software. The software is written in Ada and is checked for run-time errors using the Polyspace static analysis tool. Two types of error are detected: certain errors and possible errors.

2.6 Summary

Two types of SCA are apparent, syntactic and semantic analysis; and these are used for two purposes: demonstration of code integrity and functional correctness. Syntactic Analysis is based on the *structure* of the code and has the ability to detect control flow errors (e.g. unreachable code), data flow errors (e.g. an OUT parameter that is not always written) and coding standard violations. Additionally a syntactic technique, Information Flow analysis, can be used to support software partitioning arguments by demonstrating that sections of code are independent of each other. Syntactic Analysis can also be used to measure software and generate software metrics, though these tend to be relatively simple metrics such as McCabe's complexity metric.

Analysis of program semantics is a less commonly used, but more powerful technique. The critical difference between Semantic and Syntactic analysis is that the properties analysed by Semantic analysis require an understanding of the meaning of the software, and this enables analysis of code properties that require knowledge of the values held by program variables. For example, analysis that a divide-by-zero error will not occur requires a Semantic approach. A Semantic Analysis of the code would seem to be better suited to the demands of high-integrity software.

The sample of SCA projects reviewed above indicates that SCA is typically used for two purposes:

- **Integrity checking.** Integrity checking ensures that certain language-violations are guaranteed not to happen, i.e. that code execution will be predictable. Examples of integrity checks are:
    - Run-time errors will not occur
    - Variables are initialised before being read

- **Functional checking.** Functional checking ensures that the functional properties of the software meet their requirements. This does not include non-functional aspects such as timing properties or resource usage, but does, potentially, include checking functional safety properties.

There is therefore a wide range of static code analysis tools and techniques, with variation in rigour, objectives and cost of application. The choice facing the safety engineer is not obvious. The presence of software in his safety-related system brings with it a class of risks that, because software failures are systematic rather than random, are difficult to manage.

Static code analysis, as a powerful software analysis technique, seems to offer the potential to reduce these risks to levels commensurate with failure rates demanded of safety-related software.

3 Standards

On completion of the Darlington project, (Parnas 1990) concluded that contemporary standards were "permissive, vague and provide little guidance to either a regulator or a software developer".

Whether Parnas's comments were directly influential on standards committees or not, new software safety standards appearing in the 1990s certainly were more prescriptive and provided clear guidance to developers and regulators. Two standards are particularly relevant to this paper as they recommended the use of the SCA techniques discussed earlier: DEF STAN 00-55 and IEC 61508.

3.1 DEF STAN 00-55 (Issue 2)

The requirements for SCA given in 00-55 are as follows:

36.5.1 Static analysis comprising subset analysis, metrics analysis, control flow analysis, data use analysis, information flow analysis, semantic analysis/path function analysis and safety properties analysis; shall be performed on the whole of the source code to verify that the source code is well formed and free of anomalies that would affect the safety of the system.
36.5.2 Proof obligations shall be:
a) constructed to verify that the code is a correct refinement of the Software Design and
does nothing that is not specified;
b) discharged by means of formal argument.

There are some interesting observations to make about these requirements:

- 00-55 was widely interpreted as needing semantic analysis or path function analysis as performed by MALPAS and Spark, but not the more rigorous Compliance proofs. However, section 36.5.2, in requiring discharge by formal argument of proof obligations to verify that that the code is a correct refinement of the Software Design, would seem to require that the Compliance Analysis or interactive proof checking is performed.

- The requirements for "safety properties analysis" and analysis to be performed on "the whole of the source code" are in some sense inconsistent. Analysis, using the most rigorous techniques, of all of the source code is onerous and difficult to justify for parts of the software that may have no impact on the system safety. Analysis of "safety properties" offers the potential for a more focussed analysis. But this potential is not realised within 00-55.
- The software assurance requirements are not well integrated with the system safety management requirements given in the sister-standard DEF STAN 00-56.
- The SCA requirements are mandatory for SIL 3 and 4 software. Justifications may be given for not meeting the requirements for SIL 1 and 2. (And in practice contracts sometimes grant further exceptions.)

3.2 IEC 61508

IEC 61508 was approved after a lengthy gestation period in 1998. The standard comprises 7 parts, part 3 covering software requirements. Software functional safety requirements are identified, and a SIL identified for the software. The 61508 software safety regime can be summarised as follows:

- Software safety requirements are defined,
- Software safety requirements are validated,
- Software is designed and developed,
- The software safety is validated,
- The software is verified.

The final stage is then expanded on. Software verification is comprised of the following stages:

- Verification of software safety requirements,
- Verification of software architecture,
- Verification of software system design,
- Verification of software module design,
- Verification of code,
- Data verification,
- Software module testing,
- Software integration testing,
- Programmable electronics integration testing,
- Software safety requirements testing (software validation).

Techniques for performing these verification activities are given in Table 1, and the static analysis requirements refined in Table 2.

Technique/Measure*	SIL1	SIL2	SIL3	SIL4
1 Formal proof	-	R	R	HR
2 Probabilistic testing	-	R	R	HR
3 Static analysis	R	HR	HR	HR
4 Dynamic analysis and testing	R	HR	HR	HR
5 Software complexity metrics	R	R	R	R

Table 1

Technique/Measure*	SIL1	SIL2	SIL3	SIL4
1 Boundary value analysis	R	R	HR	HR
2 Checklists	R	R	R	R
3 Control flow analysis	R	HR	HR	HR
4 Data flow analysis	R	HR	HR	HR
5 Error guessing	R	R	R	R
6 Fagan inspections	-	R	R	HR
7 Sneak circuit analysis	-	-	R	R
8 Symbolic execution	R	R	HR	HR
9 Walk-throughs/design reviews	HR	HR	HR	HR

In the early phases of the software safety lifecycle verification is static, for example inspection, review, formal proof. When code is produced dynamic testing becomes possible. It is the combination of both types of information that is required for verification. For example code verification of a software module by static means includes such techniques as software inspections, walk-throughs, static analysis, formal proof. Code verification by dynamic means includes functional testing, white-box testing, statistical testing. It is the combination of both types of evidence that provides assurance that each software module satisfies its associated specification.

Table 2

Although these standards have been criticized as being a "nightmare to assess against" (Sampson , 2003), they do at least offer clear guidance for the software developer and address some of the problems with earlier standards that Parnas identified.

However, when looking at the philosophy behind these standards, problems again emerge. These are discussed in the following sections.

3.3 Failure rates and safety integrity levels.

The use of failure rates to determine system reliability is well established, and in IEC 61508 they are used to define Safety Integrity Levels (SILs). As we have seen, SILs are also used to categorise software components. However the use of failure rates for software is problematic, as software failures are systematic - in contrast to the random failures seen with hardware.

Techniques do exist for estimating software failure rates based on historical use data (Adelard, 2001) and work has been performed using Bayesian Belief Networks (Strigini, 1996) enabling evidence from diverse assessment activities to be combined. However, historic data cannot, obviously, be used for newly written software.

The tables given in IEC 61508 and the SIL Tailoring tables given in DEF STAN 00-55 link certain software techniques to SIL levels. However, there is little or no plausible evidence that the techniques mandated or recommended produce software of the reliability required by the SIL failure rate.

3.4 COTS

Both IEC 61508 and DEF STAN 00-55 mandate software development processes, with the implication that following the prescribed development process is essential to developing software of the required integrity. The use of Commercial off-the-shelf (COTS) software is increasingly prevalent in safety-related systems. However, the approaches prescribed by IEC 61508 and DEF STAN 00-55 cannot be used for COTS software, for which the system developer has little or no control over the development processes adopted. Furthermore, source code may not be available for COTS software, and hence many of the verification techniques recommended cannot be used.

Much of the thinking behind IEC 61508 and DEF STAN 00-55 seems to be targeted at real-time, embedded software. Modern safety-related software can be an application running on a standard desktop operating system, it can be previously developed software re-used in a new application, it can be part of a complex 'system-of-systems'. Demonstrating the safety of these types of system introduces new issues, which 61508 and 00-55 do not address.

4 The Move to Evidence Based Standards

Perhaps a change in the approach to software safety standards was inevitable sooner or later. A number of research strands looking at the issue of certifying COTS products for use in safety-related systems (Advantage Technical Consulting, 2000) , (Adelard, 2001), proposed an evidential approach.

The evidential approach emphasises the gathering and assessment of evidence that the existing *product* is safe within the proposed environment and boundaries of use. This approach may be contrasted with the approach to bespoke software safety, which combines product evidence with evidence that the *process* producing the software is sound. Two types of evidence are identified: direct evidence and backing evidence. Direct evidence is evidence that is directly concerned with demonstrating that software safety requirements have been satisfied. Backing evidence concerns secondary, but still important, aspects such as ensuring that the evidence has been generated from a system that is representative of the intended use of the software.

The first software standard to adopt an evidential approach was the Civil Aviation Authority's SW01. This is the first part of a 3 part standard (HW01, covering hardware, and SYS01 covering systems will follow). Some assumptions regarding the content of SYS01 are made by SW01, namely that "software safety requirements have been derived from a full risk and safety analysis of the system".

There are two main requirements relating to software safety:

1. ensure that arguments and evidence are available which show that the Software Safety requirements correctly state what is necessary and sufficient to achieve tolerable safety, in the system context.

2. ensure that arguments and evidence are available, which shows that the software satisfies its safety requirements.

The first of these is similar to the validation of software safety standards. No firm requirements are given for what is "tolerable safety" (although this may appear in the forthcoming SYS 01, and one can expect it to be based on standard ALARP thinking).

An important caveat is made about the standard,

> "This document does not prescribe how the assurance evidence is to be produced or its adequacy argued. International software assurance standards and guidelines, such as IEC 61508 Part 3, RTCA DO178-B/EUROCAE ED12-B, and Def Stan 00-55, when used in conjunction with this document may provide an effective way to produce timely and technically valid evidence to satisfy these assurance objectives."

How, then, can the techniques described in the process-based standards be used in an evidential approach?

SW 01 lists three types of direct evidence for safety requirements satisfaction: Testing, Design and Field Service; furthermore a number of software properties that the evidential types must be applied to are defined:

- Functional properties ("Arguments and evidence should be available that show: (a) The source code contains a correct implementation of the functional properties of the software safety requirement, either directly or by means of intermediate design notations or stages. This includes those functional properties that have been derived from non-functional software safety requirements. (b) All parameters and constants used in conjunction with the software system have been checked for correctness and internal consistency."),
- Timing properties,
- Robustness. (Arguments and evidence should be available which show that all credible modes of failure have been covered, including software failures, interface failures, power-loss and restoration, failures of linked equipment, and breaks in communication links.),
- Reliability,
- Accuracy,
- Resource usage,
- Overload tolerance.

An important observation is that evidence is required of the software *safety* requirements rather than the entire functionality. This is a long overdue development - when there is concern over the cost of demonstrating functional correctness, it seems sensible to limit the evidence gathering process to those aspects that are required do demonstrate safety.

In addition to SW01, the new draft of DEF STAN 00-56 uses an evidential approach. The software part of this standard is planned to be issued in 2004.

However, although evidence-based standards remedy some of the problems with process-based standards, an old problem identified by Parnas has been resurrected: the evidence-based standards are non-prescriptive and offer little guidance to software developers.

5 Way Forward for Static Code Analysis

The software properties list given in SW01 strike a good balance between conciseness and comprehensiveness. Static Code Analysis is ideally suited to gathering evidence for two of the properties listed: functional properties and robustness. Indeed these are the two properties identified in Section 2 that SCA has historically been used for.

Integrity (or robustness) checking ensures that the software will execute predictably. In the event of an integrity failure all software running on a processor can potentially be compromised. It is therefore necessary to provide arguments limiting the scope of integrity checking. Multiple processes, of differing SILs (including non safety-related software), can be run on the same processor. If suitable separation and non-interference arguments are used (such as the Ada Ravenscar profile), it is not necessary to perform integrity checking on all software. Having identified the scope of integrity checking, SCA tools can perform the necessary checks with a high degree of automation. Tools capable of semantic analysis of the software are required to perform a full set of the integrity checks.

Functional checking of the safety requirements ensures that the software implementation will correctly perform the software safety requirements. In order for this to be performed accurately, it is important to specify the safety requirements in a precise format. This is nothing new - a remarkable amount of work has gone into developing software specification technologies. Adapting these to produce a software safety specification offers the advantage of having a clear, well-defined safety requirement.

Once the software safety requirements have been specified, SCA tools can be used to ensure that the software implementation correctly implements the safety requirements. Software safety requirements are most often system invariants. Invariants can be translated into assertions, which can then be proved with SCA tools. However, to use SCA tools in this way requires the use of proof checking analysis techniques such as Compliance Analysis.

6 Conclusions

There are a number of fundamental problems with process-based approaches to software safety management. While these do not appear to be resulting in safety-related software products of unacceptable quality, there is a difficulty in developing cogent arguments that the risks associated with safety-related software are acceptable or tolerable.

Some of the major problems are summarised below:

- The major software safety standards promulgate a process-based approach. This approach is popular with developers as it offers clear guidance. However, it is difficult to argue that the mandated processes guarantee that the software product does not introduce intolerable risks. Moreover, process-based standards are not well suited to COTS software.
- Some new standards prescribe an evidential approach. These are better integrated with the system risks, but (currently) offer little guidance to developers and safety auditors.
- Any quantitative assessment of software safety is problematic.
- It is difficult to justify the use of SILs for software.

The move to evidence based standards offers the potential to gather evidence that the software implementation meets its safety requirements. The process to gather evidence can be more focussed that the traditional use of SCA tools, in that it is not necessary to demonstrate that all of the software requirements have been correctly implemented; only the safety requirements need to be analysed. This approach also can be used on COTS software, provided that the source code for the COTS component is available.

References

Adelard (2001) Justifying the use of software of uncertain pedigree (SOUP) in safety-related applications. Report No: CRR336 HSE Books 2001 ISBN 0 7176 2010 7.

Advantage Technical Consulting (Jan 2002) The Assessment of Software Components for Safety Applications. Technical Guidance. 21989/05-Rep-01.

Archinoff G H, Hohendorf R J, Wassyng A, Quigley B, Borsch M R (1990). Verification of the shutdown system software at the Darlington Nuclear Generating Station. International Conference on Control and Instrumentation in Nuclear Installations.

Chapman R (2000). Industrial Experience with SPARK. Praxis Critical Systems. ACM SigAda 2000.

Harrison (1999). Static Code Analysis on the C-130J Hercules Safety Critical Software. Aerosystems International, UK. Int. Systems Safety Conf. 1999

Nadjm-Tehrani S (2002). Safety Critical Systems: the role of formal techniques. Linkoping University.

Parnas D L, Asmis G J K, Kendall J D (May 1990). Reviewable development of Safety Critical Software. International Conference on Control and Instrumentation in Nuclear Installations.

Sampson C (2003). Software Dependability Issues. IMechE Workshop, "Is the Software in your Product Safe and Reliable".

Strigini L., Fenton N. (1996). Rigorously Assessing Software Reliability and Safety. ESA Software Product Assurance Workshop, Nordnjik.

UK Ministry of Defence (July 2003), Safety Management Requirements for Defence Systems, DEF STAN 00-56, Issue 3 Draft.

Ward N. J. (1993). The Rigorous Retrospective Static Analysis of the Sizewell 'B' Primary Protection System Software. TA Consultancy Services. Presented at Safecomp 93.

Estimating PLC logic program reliability

Peter G. Bishop

Centre for Software Reliability, City University
London, UK

Abstract.

In earlier research we developed a theory for predicting the reliability of conventional sequential programs based on an estimate of residual faults. This paper describes how the theory was applied to a realistic industrial example containing a known number of faults. The industrial example was implemented in a PLC application language where the program is represented by a network of logic gates (e.g. AND and OR gates). To make a residual fault estimate, our fault estimation method had to be adapted to apply to logic networks. The previous estimation method relied on a measurement of code coverage, and this had to be replaced by a measurement of logic network coverage. Several different measures of logic coverage were evaluated, including coverage of input values, output values, and input-output pair values. Using the residual fault estimate and information about the testing applied, a reliability bound was calculated and we assessed the sensitivity of the bound to changes in the operational profile.

1 Introduction

In earlier research we have developed a number of new approaches for estimating the reliability of software-based systems, namely:

1. A worst case reliability bound theory that can be applied to both continuous and demand-based systems [Bishop 1996, Bishop 2002a]. In its simplest form, this method only requires an estimate of the number of residual faults (*N*) and the number of tests (*T*), The theory predicts that after *T* test demands under a given test profile, the expected value of probability of failure on demand, E(PFD), will be bounded by:

   $E(PFD) < N/e.T$ (where e is the exponential constant, 2.718...)
2. A theory for "re-scaling" the worst case bound for new operational profiles [Bishop 2002a]. One interesting result of this theory is that a "fair" test profile (where all paths are tested equally) is not very sensitive to changes in operational profile.
3. A model that relates test coverage to the number of residual faults *N* [Bishop 2002b].

4. An extension of the reliability theory to 'fractional faults" [Bishop 2002a] where N<1. In this case it can be shown that the probability of surviving for T tests without failure is always greater than $(1 - N)$.

In this paper, we apply these methods to a realistic safety-related industrial PLC application. The research study objectives were to:

- Estimate the likelihood of residual faults in the PLC application logic
- Estimate the probability of failure per demand given the level of customer testing.

We first describe the industrial application used in this study, and then describe the approach used to estimate the number of residual faults and the probability of failure on demand for the logic example. This includes an analysis of the sensitivity of the estimate to changes in the operational profile.

2 The industrial example

The particular example used in this paper was taken from an earlier industrial research study where a control and protection system was re-implemented in software using a fail-safe PLC. The original logic controlled a mechanic device by means of motors and position sensors mounted on the device.

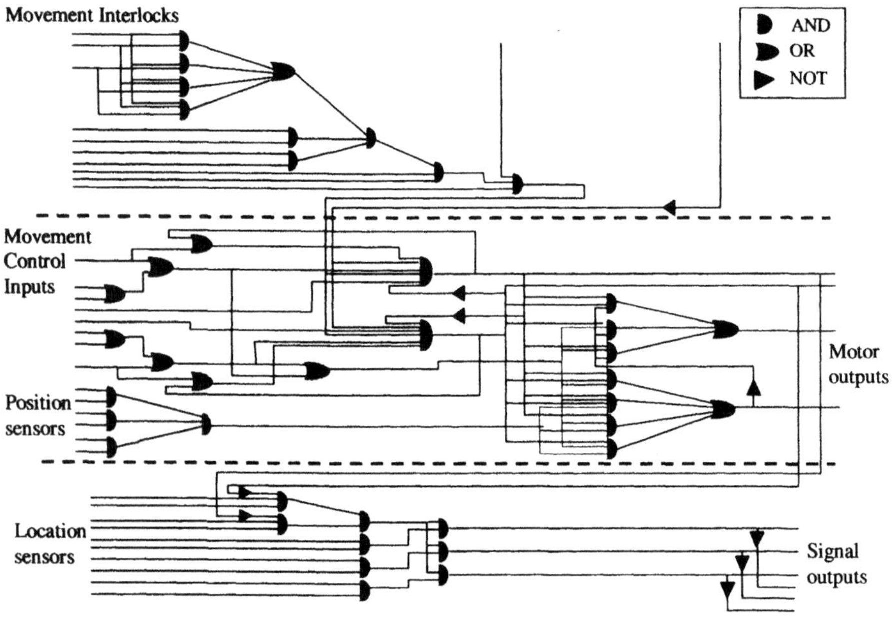

Fig. 1. Industrial control and protection logic network

As the original logic used proprietary logic circuits, the logic was translated into a logic specification. Both the customer and the PLC supplier in this study performed an independent translation. The customer translation was the basis of the oracle. The result of the customer translation is shown in the Figure 1. The logic uses AND, OR and NOT function blocks with some feedback to "latch" some of the outputs.

The supplier implementation was subjected to a range of customer acceptance tests comprising both realistic and random input sequences and compared to an independent implementation of the logic. The logic was also subjected to design review and manual testing using a switch-box.

The combination of realistic and random customer tests found 6 faults (1 interlock, 4 movement, 1 indicator signal). No faults were found in the final version, even though this was subjected to additional tests—around 10^6 random tests and around 40 000 random variations on normal operational sequences.

We used the logic specification, the set of faults found and the test specifications to evaluate our fault estimation and reliability prediction methods.

3 Residual fault estimation

To estimate the probability of failure per demand, we first require an estimate for the number of residual faults (N). A theory was developed in [Bishop 2002b] that relates the code coverage achieved with the number of residual faults. This is based on the concept of executing a "coverage element" which is some part of the code structure, e.g. a program statement, a program block between decision points, a program branch, etc. The theory assumed that:

1. A fault is equally likely to affect any one of the logic coverage elements.
2. There is a fixed probability of failure f when a coverage element is activated.
3. Faults are limited to a single coverage element.

If these assumptions apply to a measure of logic coverage, the same theory can be used. However in logic networks, assumption (3) that a fault is limited to a single coverage element may not hold. Looking at practical logic circuits, a number of different combinations of input value, K, could reveal the same fault. Applying the theory in [Bishop 2002b], it can be shown that for the case where the proportion of uncovered U decreases exponentially, the expected number of residual faults N will be:

$$N = N_0 \cdot U^f$$

Where N_0 is the number of faults prior to testing. It can be shown that if assumption 3 is violated and a fault spans K coverage elements, then:

$$N = N_0 \cdot U^{f \cdot K}.$$

Or if we define F as $f \cdot K$, this simplifies to:

$$N = N_0 \cdot U^F$$

While f cannot be greater than unity, K can exceed unity so it is possible to have $F > 1$, and in this case, the theory predicts that the proportion of faults detected can be greater than the proportion of coverage achieved.

To apply this theory to the logic example, we need to identify a suitable coverage measure for the logic, and then derive the F parameter in order to predict the number of residual faults. An ideal coverage measure, would:

- Maximise F (as this increases the rate of fault detection).
- Have a sufficiently small number of distinct coverage values to achieve 100% coverage in a reasonable time.

3.1 Logic coverage measures

The logic coverage measures that we investigated were:

1. *Input value coverage*, where all possible combinations of input values are covered.
2. *Output value coverage*, where all possible combinations of output values are covered.
3. *Input-output pair coverage*, where input values are selected such that a given input can "toggle" the state of a given output.

Complete input coverage is impractical for this logic network as it requires 2^{36} tests, but one other possibility is to use *input subset coverage*. In Figure 1, the dashed lines separate three distinct zones in the logic that only have a small number of interconnections. Given the number of inputs and interconnections between logic zones it should be possible achieve complete input coverage of the logic subsets with no more than 2^{18} tests per logic subset.

With ten binary outputs, only 2^{10} output combinations are possible for the example logic, and in practice the constraints imposed by the logic network exclude the majority of output combinations. An analysis of the intended function of the logic suggested that only 12 of the output combinations should occur. This was likely to be a relatively coarse measure for estimating the number of residual faults.

An input-output pair is defined as: *a combination of input values* $<V_1 .. V_{max}>$ *where a change of input value* V_I *will result in a change to output value* V_J. There is a strong relationship between this measure (termed I-O pair for short) and the Modified Condition Decision Coverage (MCDC) test method used in conventional programs [RTCA 93]. In MCDC testing, the values of the terms in an IF condition are selected so that a change to an individual term (e.g. from TRUE to FALSE) changes the Boolean value of the whole IF condition. The main difference when testing a logic network is that an input change can affect multiple outputs rather than the single Boolean value. To assess the coverage with respect to multiple outputs, logic coverage is measured in terms of input-output pair combinations. Note that several outputs might be toggled simultaneously when an input changes, so more than one pair might be covered by a single test. A maximum 4 input-output combinations are possible for a given *I,J* pair, so the maximum number of combinations with 36 inputs and 10 outputs is $4 \times 36 \times 10 = 1440$. In practice the number of actual combinations could much less due to constraints between inputs and outputs imposed by the logic network. The number test could be even lower as several outputs might be toggled in a single test.

As I-O pair coverage focuses on "sensitive paths", this logic coverage measure should increase the value of *F*, while the limited number tests needed to achieve coverage suggests that it should be practical to apply to realistic networks.

3.2 Coverage measurement environment

To measure the coverage achieved under testing, we implemented a coverage measurement testbed in Perl. The environment comprised several different random data test generators and predefined test sequences that could be fed into a model of the PLC logic. The logic model was instrumented to determine the coverage achieved for each measure.

The types of random test data generated were:

- uniform input random—where the probability that an input is set TRUE (pI) is 0.5 for all inputs I)
- single bit random (only one bit altered randomly on each test) but still pI = 0.5 for inputs I
- uniform output random (where the input probabilities are set so that pJ = 0.5. for all *outputs J*)

The rationale for uniform *output* probability testing is that it will maximise the output coverage growth measure, and is also likely to increase I-O pair coverage growth as there is a 50% that each output can be "toggled" by a change in an input.

In order to achieve uniform output coverage, it was necessary to devise a procedure for back-propagating assigned output probabilities to the inputs. The rules for back propagation through logic are quite simple:

AND: $p_{in} = p_{out}^{1/n}$
OR: $p_{in} = 1-(1-p_{out})^{1/n}$
NOT: $p_{in} = (1-p_{out})$

where:

n is the number of inputs to the logic gate,
p_x is the probability of TRUE value on link *x*,

This is illustrated below for a simple single-output network, where a *p* value of 0.5. is back propagated to the inputs.

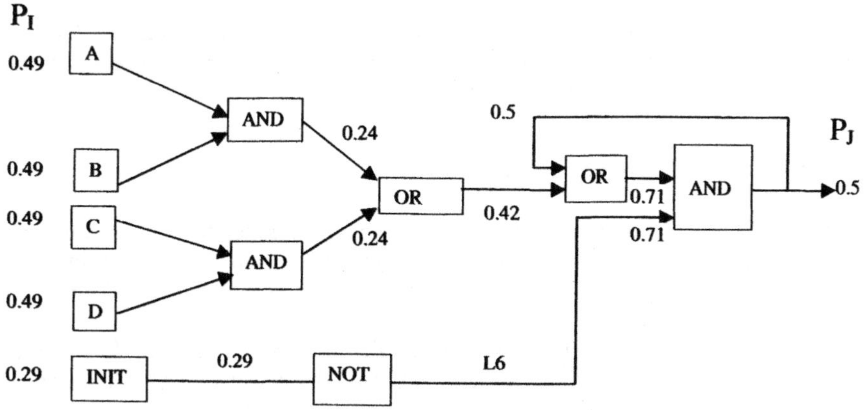

Fig. 2. Illustration of back propagation of probabilities

In practice, back propagation is constrained by network junctions and feedback loops. For example in Figure 2 above, the feedback loop forces one input to the OR gate to be 0.5. When one input probability is constrained p^*_{in}, it can be shown that he probability for the remaining inputs p_{in} is:

$$p_{in} = 1-((1-p_{out})/(1-p^*_{in}))^{1/n-1}$$

hence for p^*_{in}=0.5, p_{out}=0.71 and n=2, we obtain 0.42 for the other input. When back-propagation was applied to the actual network of 36 inputs, 10 outputs, the interconnection constraints meant that the "ideal value" of $p_J = 0.5$ could not be achieved (as negative values for p_{in} are derived during back propagation). Compromise values for the output probabilities (typically of the order of 0.3) were chosen instead to obtain valid input probabilities. The distribution of input probabilities to achieve near uniform output probabilities is shown in Figure 3. It is clear that the input probabilities can be quite extreme (close to zero or 1).

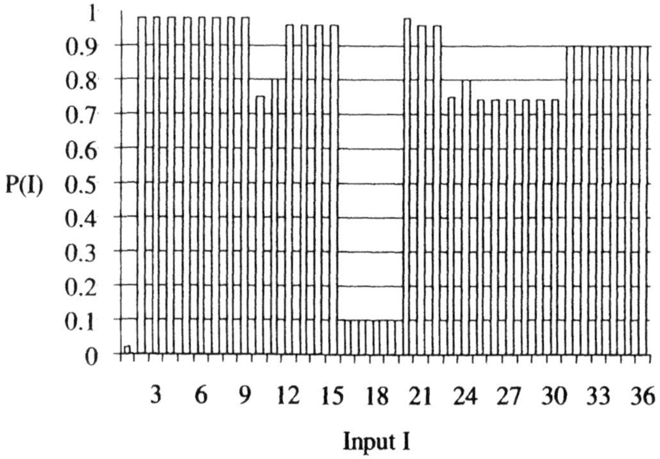

Fig. 3. Input probability distribution for near-uniform output probability

We can view this input distribution as the probabilistic equivalent of an MCDC test pattern as it maximises the chance of specifying inputs patterns where an output will change if a single input bit is changed.

3.3 Assessment of coverage growth

Using the PLC logic simulator, the coverage achieved under MCDC random and uniform random tests was measured. We also replicated some of the customer test where data was available. The coverage growth achieved under MCDC random testing is shown in Figure 4.

As can be seen from Figure 4, under MCDC random testing, full output coverage is achieved in around 100 tests, and full I-O pair coverage is reached in around 3000 tests (equivalent to 236 different I-O values). By contrast, input coverage of the logic sub-nets grows more slowly (as expected, as there are more input coverage value combinations).

The logic was retested with a uniform random input profile, the most striking changes in the growth curves were:

- Output and I-O pair coverage was much slower—the output coverage was only achieved after 106 tests, and I-O pair coverage only reached 84% after 106 tests.
- Input coverage was better—100% coverage of the interlock logic is achieved after 10^5 tests

This is consistent with expectations as uniform random input testing should maximise growth of the input coverage measure.

Coverage was also measured for the original customer "mixed" test data file of 10 000 tests comprising:

- operational sequence (159 tests)
- static interlock test data (28 test)
- dynamic interlock test data (243 tests)
- uniform random test input data (9514 tests)

The operational sequence tests proved to be the most effective means for achieving IO pair and output-coverage growth.

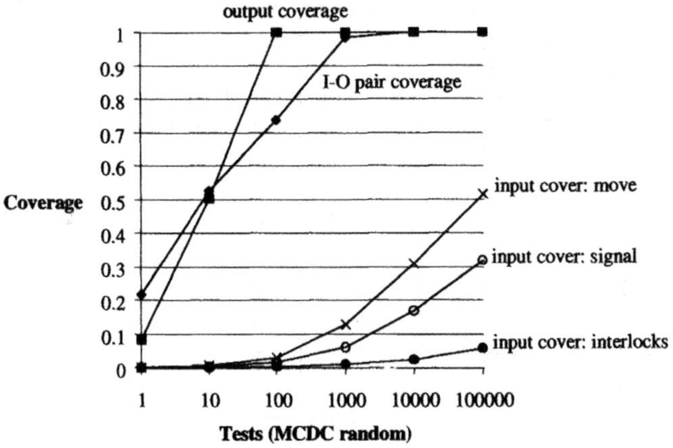

Fig. 4. Growth in coverage measure (MCDC random test profile)

3.4 Relating coverage measures to faults detected

We designed the PLC logic simulator so that the individual logic faults could be switched on and compared to the behaviour of the correct logic network when a test is applied. The cumulative fault detection counts under different test profiles are summarised in Table 1.

Table 1. Faults detected versus number of tests (different test methods)

Number of tests	Faults detected		
	MCDC test	Uniform Random	Customer tests
10	4	0	-
100	5	0	-
486	6	0	4 (ops)
1000	6	0	-
3000	6	1	-
9514	6	4	5 (random)
10000	6	4	6 (mixed)
100000	6	6	-
1000000	6	6	-

The MCDC random testing seems to be highly effective with 4 faults detected in the first 10 tests, and all 6 faults detected after 486 tests.

We then investigated whether the fault detection performance was correlated to the various types of coverage measure. Ranking input coverage against detected faults produced little obvious correlation. For example, input coverage growth was better with uniform random than MCDC, but a fault in the interlock logic was revealed 81 times in 10 000 MCDC tests and zero times in 10 000 uniform random tests.

Output coverage exhibited a better ordering between coverage measure and detected faults with different test profiles, but the coverage hit 100% before all faults were revealed. We concluded that output coverage was too coarse for predicting residual faults with any accuracy.

The best ordering was found between faults detected and I-O pair coverage. This is shown in Figure 5 above, together with the fault detection curve predicted by the model described in Section 3.

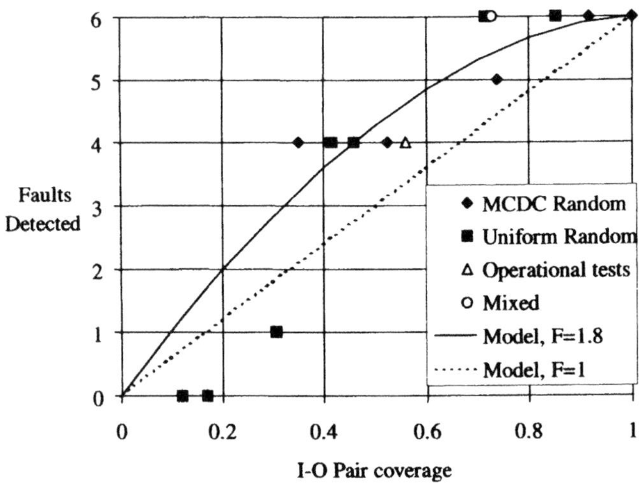

Fig. 5. I-O pair coverage vs. faults detected: comparison with model

While a simple linear model (F=1) is a reasonable approximation to the data for this logic network, a better fit at high coverage is achieved with a model parameter of F=1.8. This is in line with expectations, as we know that F can be greater than unity if a fault spans several coverage items. It also means that complete fault detection is likely to be achieved at lower coverage values. For example the uniform random tests found all 6 faults with only 70% I-O pair coverage.

3.5 Residual fault estimation using the model

Having parameterised the model it is possible to convert a coverage measure into a residual fault estimate. Using the logic simulator we measured the coverage achieved using the customer tests that were applied to the Intermediate version of PLC logic implementation. The customer tests are summarised in Table 2.

Table 2. Customer tests applied to Intermediate version of PLC logic

Test Type	Number of tests
Random Single Input Tests:	60 000
Total Random Tests:	160 000
Plant Operational Sequence Tests:	159

The coverage achieved by applying around 220 000 customer tests was 195 I-O pairs, i.e. the fraction of I-O pair coverage is 195/236, hence C = 0.826, so the fraction that is not covered is U = 0.174. According to our coverage model, the fraction of undetected faults is:

$$N/N_0 = U^F$$
$$= 0.174^{1.8}$$

Therefore:

$$N/N_0 = 0.042$$

Assuming that the 6 faults found represent N_0, the estimate for residual faults is:

$$N = 0.26$$

This is an example of the 'fractional fault" case that was examined in [Bishop 2002a]. The result could also be interpreted as saying that there is at least a 74% chance of zero faults in the logic. This result should be reasonably conservative as some customer tests were not available for inclusion in the coverage measurement. A purely linear model would have predicted N=1.

By comparison, 3000 MCDC random tests achieved full coverage of 236 I-O pairs, which would result in an estimate that N is effectively zero.

4 Application of the reliability theory

4.1 Scaling reliability bounds

Our worst case bound research [Bishop 2002a] suggests that it is possible to take a reliability bound for a given test profile and scale it up for a different operational profile. The basic assumption in this theory is that, for a fault is associated with coverage element i (e.g. a particular output value), the failure rate is proportional to the element execution rate, $X(i)$. If the logic is tested under one profile and operated under another profile, it follows that the failure rate is scaled by:

$$S(i) = X_{op}(i)/X_{test}(i)$$

Of course we do not know which coverage element i a fault is located in, so the mean scale factor S is the average of $S(i)$ over all coverage elements. The theory in [Bishop 2002a] shows that the reliability bound is also scaled by S when the profile changes, i.e.:

$$E(PFD) < S{\cdot}N / e{\cdot}T$$

In addition the research showed that a 'fair" testing profile might be derived where the reliability bound is relatively insensitive to changes in the operational profile.

If the MCDC random test profile is close to a 'fair test", all coverage elements should be evenly tested. This is quite difficult to assess for I-O pair coverage where there could be constraints between I-O pair values, but analysis of the coarser output value coverage yields the following results.

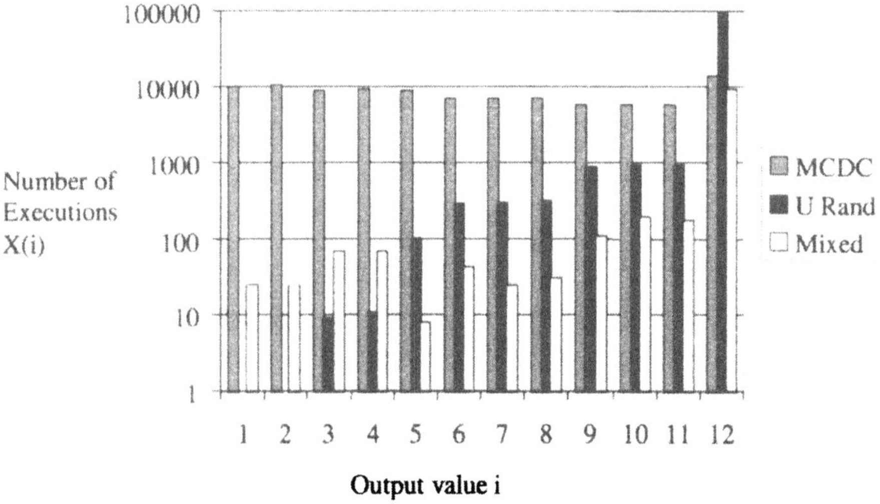

Fig. 6. Execution of output coverage elements

It is clear the MCDC random testing produces an output coverage that is close to an even distribution, while the uniform random input tests produce the most asymmetrical distribution. Assuming that a fault only affects one coverage element (a reasonable assumption for the relatively coarse output coverage), the re-scaling theory developed in [Bishop 2002a] can be applied.

Note the calculation of *S* assumes that the actual faults present in the logic are equally likely to affect any coverage element. A more pessimistic scale factor would be the mean of the *N* largest value of *S(i)* (as there are *N* faults). Similarly a more optimistic scale factor would be the mean of the *N* smallest *S*(i) values. Table 3 shows the scale factors derived from the execution rate distributions in Figure 6.

Table 3. Predicted scale-up factors for different test and operational profiles

Test Profile	Operational Profile	Scale factor		
		Min N	Mean	Max N
Uniform random	MCDC random	10.8	1893	3775
MCDC random	Uniform random	0.03	0.64	1.3
Mixed	MCDC random	6.4	25	44
MCDC random	Mixed	0.009	0.62	1.2

It can be seen that the theory predicts that the use of "unbalanced" profiles for testing (e.g. uniform random or mixed) will lead to very large increases in the calculated PFD bound when a dissimilar profile is used in operation. By contrast the PFD bound derived using a balanced test profile might even decrease when a dissimilar profile is used in operation.

If we take the case where uniform random data is used in testing, the theory in [Bishop 2002a] predicts that the bound on the PFD under this profile after T tests will be:

$E(PFD) \le N / e.T$

If the operational profile is MCDC random was used in operation the theory predicts that the operational PFD will lie within a rescaled bound of:

$E(PFD) \le 3775.N / e.T$

where 3775 in the pessimistic scale factor in Table 3. If the situation is reversed and MCDC is used in testing and uniform random is used in operation, the theory predicts that the operational PFD will lie within a rescaled bound of:

$E(PFD) \le 1.3.N / e.T$

where again we take the more pessimistic scale factor in Table 2.

4.2 Evaluation of the reliability bound predictions

We evaluated the bound predictions using the known faults present in the logic. The failure probability per test, *p(n)* of each fault under different profiles were measured using the logic simulator with different faults enabled. The failure rates under different profiles are shown in Table 4.

Table 4. Failure probability per test, *p(n)* under different test profiles

Fault *n*	*p(n)* MCDC	*p(n)* Uniform
1	0.008	0.00020
2	0.40	0.00014
3	0.40	0.00015
4	0.39	0.00004
5	0.39	0.00021
6	0.30	0.00002

It can be seen that MCDC random testing achieves the highest failure probabilities for each fault.

Using the measured failure probabilities for the logic defects in Table 4, reliability growth was modelled under one profile (to represent testing), then the reliability was re-computed under a different operational profile. The failure probability after *T* tests followed by a switch to new operational profile, is determined by the equation:

$$E(PFD') = \sum p'(n) \cdot (1-p(n))^T$$

where $p'(n)$ is the failure probability of fault *n* under the new operational profile, and $p(n)$ is the failure probability of fault *n* under the original test profile.

Figure 7 shows the reliability growth curves for these faults under the original test profile (uniform random) and the operational profile (MCDC random), together with the original and scaled bound predictions. Note that the bound predictions are not based on a knowledge of the actual logic defect failure probabilities, but only on an estimate of the number of faults (assumed to be N=6) and the execution rates of the output coverage elements under different test profiles. It can be seen that the PFD of the actual faults are below the predicted worst case bound, and the PFD changes by three orders of magnitude when the profile is changed. The PFD would have exceeded a bound based on the mean scale factor (S=1893), so the faults may disproportionately affect coverage elements with the largest S(i) values.

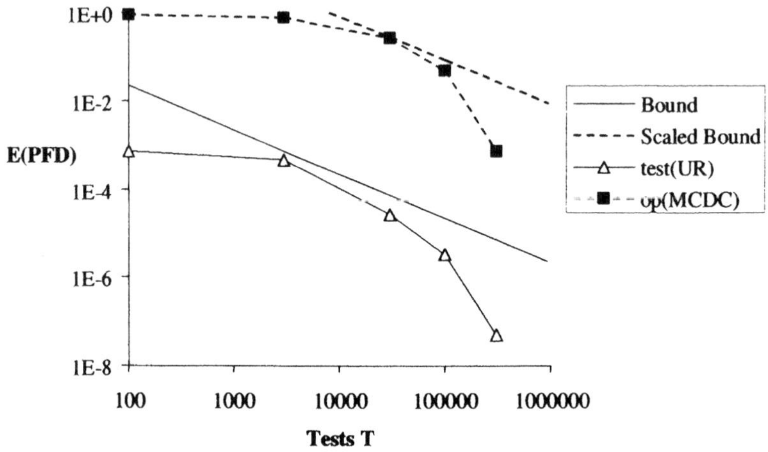

Fig. 7. Re-scaled worst case bound for a new operational profile

It is clear that the reliability prediction is highly sensitive to the test profile used. We repeated the evaluation for the converse case (MCDC random data during testing, uniform random for operation). The results are shown in Figure 8.

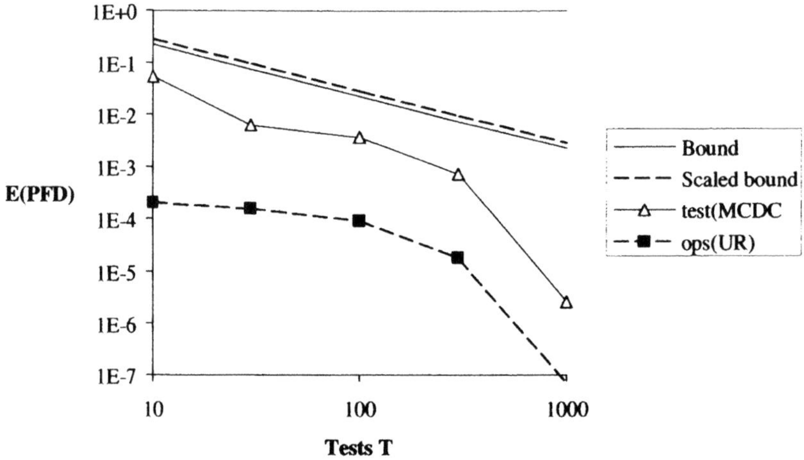

Fig. 8. Rescaled bound: after 'fair" MCDC testing

It can be seen that the PFDs lie below their respective bounds. In this case, we predicted that the scale factor for the bound would be no more than 1.3, but in fact the actual PFD *decreased* by two orders of magnitude (i.e. the scaling is less than unity). This is close to the more optimistic predicted scale factor of 0.03 (see Table 3).

4.3 Comparison of reliability predictions

From an analysis of $2.2 \cdot 10^5$ customer tests we estimated that the number of residual faults was N=0.26. What does this imply for the future reliability of the logic software? We compared the predictions of the black box Bayesian method of [Littlewood 1993] and the "worst case" reliability model presented in [Bishop 2002a].

The black box Bayesian method is based on the last failure free interval during testing T_0, so this interval will be less than $2.2 \cdot 10^5$ as 6 faults were found during testing, but we will assume $T_0 = 2.2 \cdot 10^5$ to give a best case prediction.

The worst case reliability function [Bishop 2002a] has similar in shape to the Bayesian reliability function where the probability of operating without failure $R(t|T)$ decreases almost inversely with the number of tests t. We have two ways of using the worst case theory:—using the initial value of N=6 and $T=2.2 \cdot 10^5$, or the final fractional value of N=0.26, but claim no credit for prior testing (T=0). As mentioned earlier, when the number of residual faults N is fractional, the probability of survival is asymptotic to 1–N; for the case where N=0.26, the asymptote is 74%. The reliability predictions are shown Figure 9.

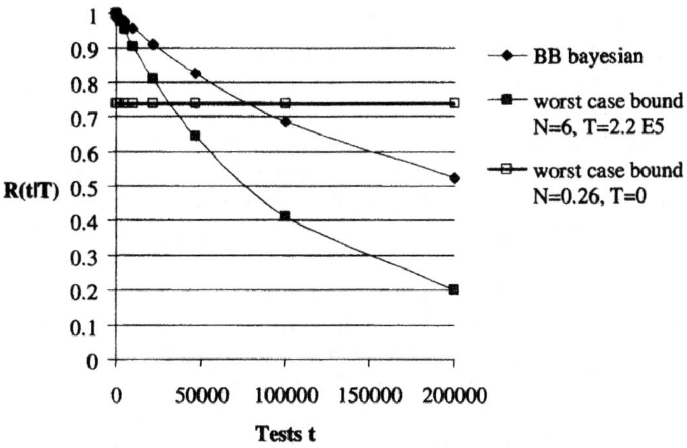

Fig. 9. Comparison of reliability predictions

The N=6 worst case reliability model predicts a 41% chance of surviving a further $t=10^5$ test demands, while the Bayesian method predicts a survival probability of at most 69%. However, using the N=0.26 case, the survival probability would be 74% and would remain so for any number of further demands. As the worst case functions are both bounds, it is legitimate to take the maximum of the two bound predictions. In addition, the N=0.26 prediction is unaffected by changes in operational profile. For the "mixed" test profile, scale factors of 25 in the failure rates are possible in operation (see Table 3), which would reduce the predicted survival times by a factor of 25—but this rescaling makes no difference to an asymptote.

The apparently optimistic prediction of the worst case reliability theory seems to be supported by the fact that no faults were detected by the customer when the PLC logic was retested with more than 10^6 tests after fixing the detected faults.

5 Summary and conclusions

Previously, the reliability theory had only been applied to conventional program code. However it proved to be fairly easy to adapt the theory to PLC logic networks. Minor extensions to the coverage growth theory were needed, and we also needed to identify suitable coverage measures for logic rather than conventional code. Once this had been done, it was possible to use:

1. coverage growth theory for fault estimation
2. re-scaling theory to adjust a PFD bound for a new operational profile
3. the concept of a balanced test profile to derive a PFD that is insensitive to changes in operational profile
4. the concept of a 'fractional residual fault" to give less pessimistic reliability estimates than other reliability models.

It should be noted that the reliability estimation described in this paper only applies to faults in the application logic. Other faults in the PLC (e.g. in the firmware) have to be addressed separately.

Output and I-O pair coverage measures were found to correlate with detected faults and both measures might be applicable for estimating faults in other logic networks.

There is a non linear relationship between coverage and faults. A coverage growth model can be fitted to the observed data to estimate residual faults, but it is probably conservative to assume a linear relationship between coverage and faults found, and then devise a test strategy that maximises the coverage (like MCDC random testing).

We found that the "statistical MCDC" test method was the most effective for achieving high I-O pair and output coverage, while uniform random input testing was the least effective. The "statistical MCDC" method also appears to be a 'fair" test profile as there is fairly balanced coverage of output values.

The scaling theory in [Bishop 2002a] predicts that a PFD bound estimate derived using a test profile that gives balanced coverage will be insensitive to changes in operational profile. This prediction was supported by simulated reliability growth calculations applied to the known faults using balanced and unbalanced tests.

We conclude that:

1. Coverage analysis theory can be used to estimate the number of residual faults in logic networks.
2. Worst case bound and rescaling theory can be used to support software reliability claims.
3. MCDC random tests should be used more widely for testing logic systems both for detecting faults and for long term reliability testing.

6 Acknowledgements

This work was funded by the UK (Nuclear) Industrial Management Committee (IMC) Nuclear Safety Research Programme under British Energy Generation UK contract 40029264 with contributions from British Nuclear Fuels plc, British Energy Ltd and British Energy Group UK Ltd. The paper reporting the research was produced under the EPSRC research interdisciplinary programme on dependability (DIRC).

References

[Bishop 1996] P.G. Bishop and R.E. Bloomfield, "A Conservative Theory for Long-Term Reliability Growth Prediction", IEEE Trans. Reliability, vol. 45, no. 4, pp. 550-560, Dec. 1996.

[Bishop 2002a] P.G. Bishop, "Rescaling Reliability Bounds for a New Operational Profile", International Symposium on Software Testing and Analysis (ISSTA 2002), (Phyllis G. Frankl, Ed.), vol. 27 (4), pp. 180-190, ACM Software Engineering Notes, Rome, Italy, 22-24 July, 2002

[Bishop 2002b] P.G. Bishop, 'Estimating Residual Faults from Code Coverage", Safecomp 2002, pp. 163-174,Catania, Italy, 10-13 Sep. 2002.

[Littlewood 1993] B. Littlewood and L. Strigini, "Assessment of ultra-high dependability for software-based systems", CACM, vol. 36, no. 11, pp.69-80, 1993.

[RTCA 1993] RTCA DO-178B , 'Software considerations in airborne systems and equipment certification", 1993.

A Systematic Framework for the Assessment of Operating Systems

Philippa Conmy, Steven Crook-Dawkins,
Origin Consulting (York) Limited,
Innovation Centre,
Innovation Way,
Heslington,
York, England
YO10 5DG, UK

(philippa|steve)@origin-consulting.com

Abstract

Operating systems (OS) are common to almost all computing platforms including those used in safety related systems (SRS). OS are commercial components and, as with other "off the shelf" components, there can be significant difficulties in assessing their dependability cost-effectively. An OS provides a broad range of support services to application software, hence it cannot easily be assessed independently of the services of the application it supports. Also many of the functions of the OS are at a low level, but are so influential to the operation of the system that they become an intrinsic aspect of deployment risk for those systems. Thus we require a specialised approach to assessing dependability of OS.

This paper considers the use of an assessment framework for OS from the air traffic control domain. Our findings are based on a case study that applied the assessment framework to a simple OS: the L4 micro-kernel. The paper considers the issues in, and potential benefits of, using such a general framework.

1. Introduction

The assessment of operating systems (OS) is not a new problem, but it is becoming more important as more safety related systems (SRS) employ commercial OS. There are many difficulties of assessing OS for use in SRS. The OS are becoming increasingly complex and configurable. Intellectual Property Rights (IPR) may block access to the information required for assessment. It is difficult to determine the dependencies between the OS and any given application, and even more difficult to assess the dependability independently of an application. Thus there is a need for a cost-effective and systematic assessment approach to address these issues. More specifically, an approach is required which discharges at least two basic requirements, i.e. it should:

- Provide structured evidence and argument that support assessors in forming a view as to whether a specific OS can be used in a specific application.
- Provide a convenient and objective means of comparing different OS packages.

In addition, there are several secondary requirements which reflect the practical application of such a framework, e.g. it should:

- Assume no prior knowledge (on the part of the assessors) of the OS design.
- Support assessment of existing OS systems, including off-the-shelf (OTS) products.
- Enable assessment of all kinds of OS (e.g. general purpose, real-time, micro-kernels etc.).

The approach adopted here provides an assessment framework based around analysis of low-level functions within the OS and the (high-level) services provided to the SRS, by the OS. We believe that this goes some way towards meeting the above requirements and providing a systematic approach to reaching decisions as to whether an OS could validly be deployed in support of a specific SRS. This approach ensures that the role of assessor for the OS can be independent of, and different in scope from, that of the systems safety assessor. The crux is to ensure that the OS assessment identifies and provides evidence about properties of the services and functions in such a way that this information can be used in the assessment of the application, without the application assessor needing to understand details of the OS (and vice versa).

This paper considers an approach was developed for the UK Health and Safety Executive (HSE) and the Civil Aviation (CAA) Safety Regulation Group (SRG) based on the original work of Conmy and McDermid (Conmy 2001). The rest of this paper is structured into five sections. Section 2 gives the background to the study; section 3 describes the specific approach we used; section 4 describes our experience in using the approach; section 5 gives our recommendations and section 6 records our conclusions.

2. Background

This section describes three key areas that form the background of this study: argument context, regulatory framework and the rationale for an assessment framework. The argument context is intended to identify what it is sensible to claim about an OS in isolation. The discussion of regulatory framework identifies how such claims might be used in support of regulatory decisions. This discussion is used as a foundation for explaining the nature of the assessment framework.

Argument context

It is not meaningful to claim that an OS is safe (or unsafe) in isolation. Properties of the OS are not safety properties unless they are assessed in the context of a specific SRS. Thus, a general OS assessment framework will focus on the degree of confidence we have that an OS will comply with its published specification, and the effects that may arise due to credible deviations from its specification. (For generality, the discussion assumes that an existing "off the shelf" (OTS) OS is being assessed.) This has two implications on the assessment process used:

1. "Pragmatic" analysis. The assessment will tend to be constrained by the evidence that can be collected from existing specifications, records of usage, and analysis based on any visible design data.
2. Traceability. The assessment must allow tracing of assessment evidence, and knowledge about faults, to individual OS functions and services.

The consequence is that we cannot make claims about an OS as a whole, but about the individual functions and services it offers. However this means that the assessment of an OS may become diffuse because arguments and evidence about services the OS offers are developed almost in isolation from each other. This could lead to inefficient arguments and make the assessment expensive. Thus a framework is needed that makes explicit the role of each service and function and allows evidence about low-level functions to be gathered independently. This evidence can later be combined to provide evidence about the higher level services. To do this requires some level of design knowledge.

Regulatory framework

Our study investigates the use of OS within the Air Traffic Service (ATS) domain, where the standard "Regulatory Objectives for Software Safety Assurance in ATS Equipment " (SW01) is used. SWO1 is part of CAP 670 (CAA 2003) and establishes the main concepts required "to ensure that the risks associated with deploying any software in a safety related ATS have been reduced to a tolerable level". These concepts, and the relations between them, are illustrated in Figure 1.

The safety objectives are the key requirements to be discharged and are the only regulatory (normative) part of the standard. All software components must be shown to comply with these safety objectives through appeal to arguments and evidence. "Software components" are implemented software units that provide specific functionality to the SRS. The criticality of each safety objective for each software component will influence the degree of rigour and the techniques appropriate for gathering evidence. Finally, several aspects of software behaviour must be made explicit to ensure the argument addresses both functional and non-functional aspects. Each concept is described in more detail below.

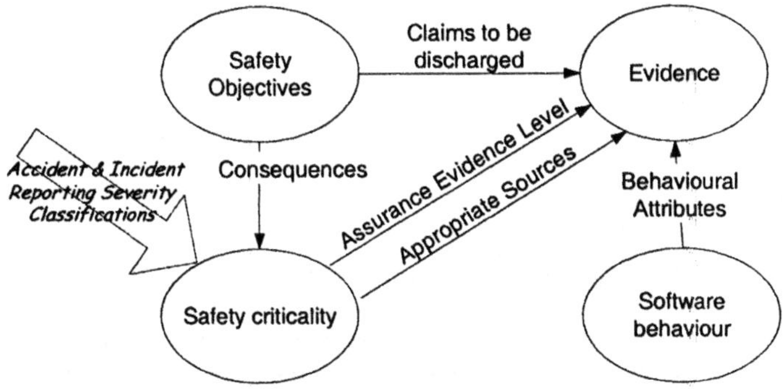

Figure 1: SW01 Concepts

Safety Objectives

Five safety objectives are prescribed:

1. Requirements validity.
2. Requirements satisfaction.
3. Traceability.
4. Non-interference with safety functions.
5. Configuration and consistency.

These objectives are derived from the perspective the software itself has no intrinsic safety properties, and must be shown to do what it was required to do, in the context of the system safety assessment. This is a consistent with the arguments made above about an OS not having specific safety properties.

Criticality

SW01 states that: "The safety criticality of the software safety requirement is expressed as an index in the range 1 to 5. This index is referred to as the Assurance Evidence Level (AEL). The AEL determines the minimum set of assurance evidence that is required to be available to the regulator for a given software safety requirement for any system proposed for approval. One interpretation of the index provided by SW01 is reproduced below (CAA 2003):

AEL	Severity
1	No immediate effect on safety.
2	Slight reduction in safety margins.
3	Major reduction in safety margins.
4	Large reduction in safety margins.
5	Complete loss of safety margins.

The main application of the AELs is to give guidance for the characteristics and rigour of evidence gathered to show that the safety objectives have been met.

Software behaviour

SW01 also defines attributes of behaviour as follows:

1. Function.
2. Timing.
3. Resource usage.
4. Overload tolerance.
5. Robustness.
6. Accuracy.
7. Reliability.

These are used to structure the evidence, particularly when arguing that supplies software satisfies requirements. SW01 defines sources of evidence that are appropriate to address each specific attribute. Thus the specific attributes and non-functional properties of the software are made explicit, helping to ensure arguments are consistent and unambiguous.

Evidence

In the guidance section, SW01 describes many aspects of the evidence required to discharge each safety objective. Two distinct types of evidence are defined:

- Direct Evidence: an attribute or property of the software, derived by an objective means, e.g. measurement or analysis.
- Backing Evidence: shows that the direct evidence is both credible and soundly based, e.g. configuration management records.

Three sources of evidence are defined:

- Testing.
- Field service experience.
- Analysis.

The guidance section of SW01 uses these categories to provide specific guidance on the evidence required to meet each objective.

Summary

In applying the standard the worst-case consequence of a software component failing to meet a specific safety objective is identified, and an AEL assigned. The standard then specifies the evidence required for that combination of safety objective and AEL. In most cases, this involves showing sufficient rigour of direct and backing evidence from test, analysis and field service. Showing requirements satisfaction always requires evidence to be gathered about software behaviour.

Rationale for an Assessment framework

Before describing the specific assessment framework and the case study, it is useful to define the rationale behind adopting an assessment framework to provide an effective response to the regulations. Later in the paper, in recommendations, we re-consider this rationale in the light of the case study.

Focussing on services

Focusing on services enables a consistent model of the OS to be produced that allows us to compare OS products.

Focus on evidence during certification.

Focusing on evidence, rather than prescribing development processes, allows us to assess OTS products directly in the same framework as bespoke software.

Commercial Issues.

OS developers work to the requirements of the general consumer market with little consideration for the special needs of the (small) SRS market. The provision of a simple framework that is easy to apply to OS may increase the willingness of suppliers to provide information.

Defining criticality as a requirement on evidence for individual services may be more intelligible and acceptable to OS suppliers than the concept of a Safety Integrity Level (SIL) applied to the OS as a whole.

Scalability

Many OS are large and complex. A framework that analyses them on the basis of functions and services may reduce the burden of assessment, as many of the more sophisticated services will not be used in an SRS.

3. Definition of an Assessment Framework

The approach put forward consists of a generic framework, an assessment process plus an approach to documentation and reporting.

General framework

The general framework for OS assessment is based on the work of Conmy and McDermid (Conmy 2001) and is shown in Figure 2 below. This is a general framework to be used to structure the assessment of specific OS products.

Each OS function (shown as lozenges in Figure 2) depends on a number of basic services (shown as rectangles). Services are groups of system API calls and low-level resource management routines. Some services are used by all functions; these are the four at the bottom of the figure. It is important that this framework be subjected to peer review as a generic framework representing the basic infrastructure of a general OS. Using the framework, a taxonomy of services for a specific OS can be built up by listing all the system calls and support routines that make up that service.

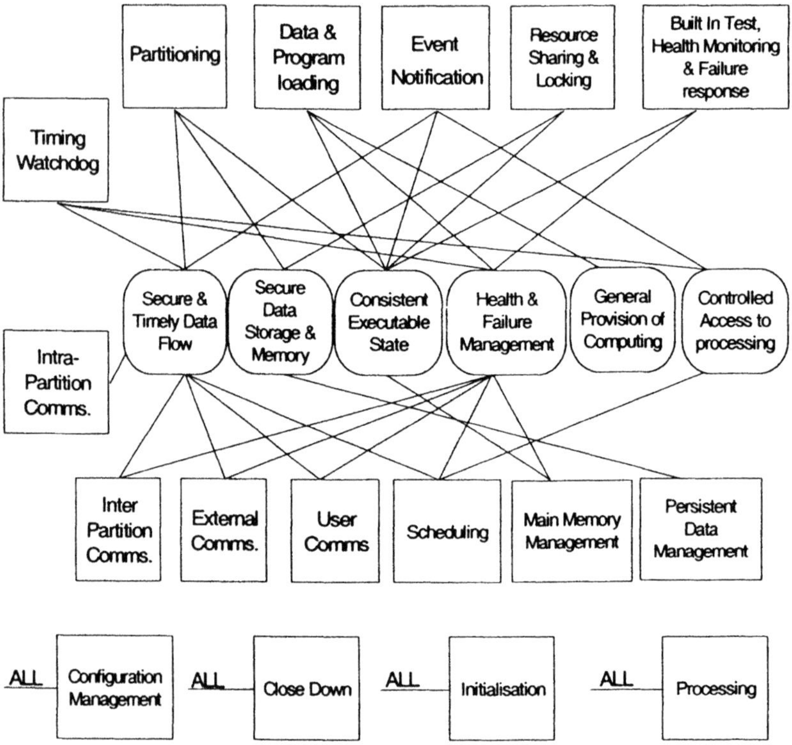

Figure 2: Framework of Operating System Functions and Services

The OS assessment process consists of several steps

1. Populate the framework by taking each service of the OS and listing the specific system calls used by the OS to implement that service.
2. Create a mapping between these services and the functions. (One possible mapping is shown in Figure 2.)
3. Assess the AEL for each service, based on the evidence available for that service against each of the five safety objectives.

4. Derive an AEL for each OS function by taking the lowest AEL of each of the contributing services.
5. Create a table summarising the information and providing a taxonomy of functions, services and references to evidence.

Once the OS assessment has been produced, it can be employed in the assessment of the suitability of the OS for a specific SRS application. More detailed discussion of the assessment of an SRS is outside the scope of this paper. For more information see (Kelly, 1998).

Documentation and Reporting

Results are communicated through two main reports :

- Part 1: The summary assessment report. This will contain a concise explanation of the assessment and record the evaluated AEL of the functions and services. This report will also give an opinion on the likely maximum application AEL the OS could support.

- Part 2: The detailed assessment report. This will contain detailed information in support of, and justifying the conclusions made in, the Part 1 report. This part will also contain additional context about the specific OS under assessment.

These reports are intended to be used by the developer and assessor of any SRS intending to use the OS. Suppliers of OS products might also use the reports to assess the suitability and risks of deploying products in support of specific systems or use the findings to guide further development.

The next section records our experience of applying these concepts in practice.

4. Undertaking the assessment

The study on which this paper is based involved an assessment of the Fiasco L4 micro-kernel (Au 1999, Tews 2001) referred to below as "L4". L4 is not, of itself, an operating system, but offers the basic services that would be required to support applications. A characteristic of such a micro-kernel is to segregate services which need to run in privileged mode from applications and broader OS services that only need to operate in user mode. This makes the micro-kernel more secure, smaller and easier to reason about than a full OS. It is also likely that many SRS will use a micro-kernel, so L4 provides a good basis for experimentation and evaluating the assessment framework.

The remainder of this section describes our experience in following the process described in the previous section.

Populating the Framework

Populate the framework with the specific system calls used

L4 supports seven main system calls as described in Table 1 below. Using descriptions and examples from the L4 documentation it was possible to establish which of the services in the framework each system call would contribute to. The results are shown in Figure 3.

System Call	Function / brief description
task_new	Task creation and deletion.
Thread_switch	Releases the processor so that another thread can consume processor time.
lthread_ex_regs	Reads and writes register values of a thread in the current task.
lthread_schedule	Used to set the priority, timeslice length, and external pre-empter of other threads. Also provides thread states can be retrieved through the appropriate scheduler.
IPC	Inter Process Communication, unbuffered and synchronous. Utilises blocking and timeout facilities. Short messages of 8-bytes can be implemented using registers alone.
id_nearest	If *nil* is specified, then this call returns the ID of the current thread, otherwise it returns the ID of the nearest partner engaged when sending a message to a specified location. Its primary function is to provide ID's during IPC or setting up registers for a thread.
Fpage_unmap	The specified fpage is unmapped from all address space into which the invoker mapped it.

Table 1: L4 system calls

Create a mapping between these services and the functions

The mappings from services to functions used in Figure 3 are derived from (Pierce 2003). These mappings may not be appropriate for other micro-kernels or full OS, as they are specific to the OS design.

Assessing the AEL for each service

Assigning AELs to system calls

The assessments were made by reviewing the evidence available for each system call from L4 documentation and any public domain literature, then by comparing this evidence against the requirements of SW01 for each safety objective. SW01 provides extensive guidance on requirements satisfaction (safety objective 2) and this tended to dominate our assessment of each L4 system call. SW01 contains guidance on the rigour of evidence required at each AEL in the form of tables. The tables detail the rigour of evidence expected from testing, field service evidence and analysis for each AEL level.

Every L4 system call listed in Table 1 was assigned an AEL of 2. The reason that system calls could not be shown to meet an AEL of 3 is that much of the evidence available is based on ad-hoc testing and field service evidence, with little analysis. Guidance within SW01 suggests that at AEL 3 and above the testing process itself should be verified.

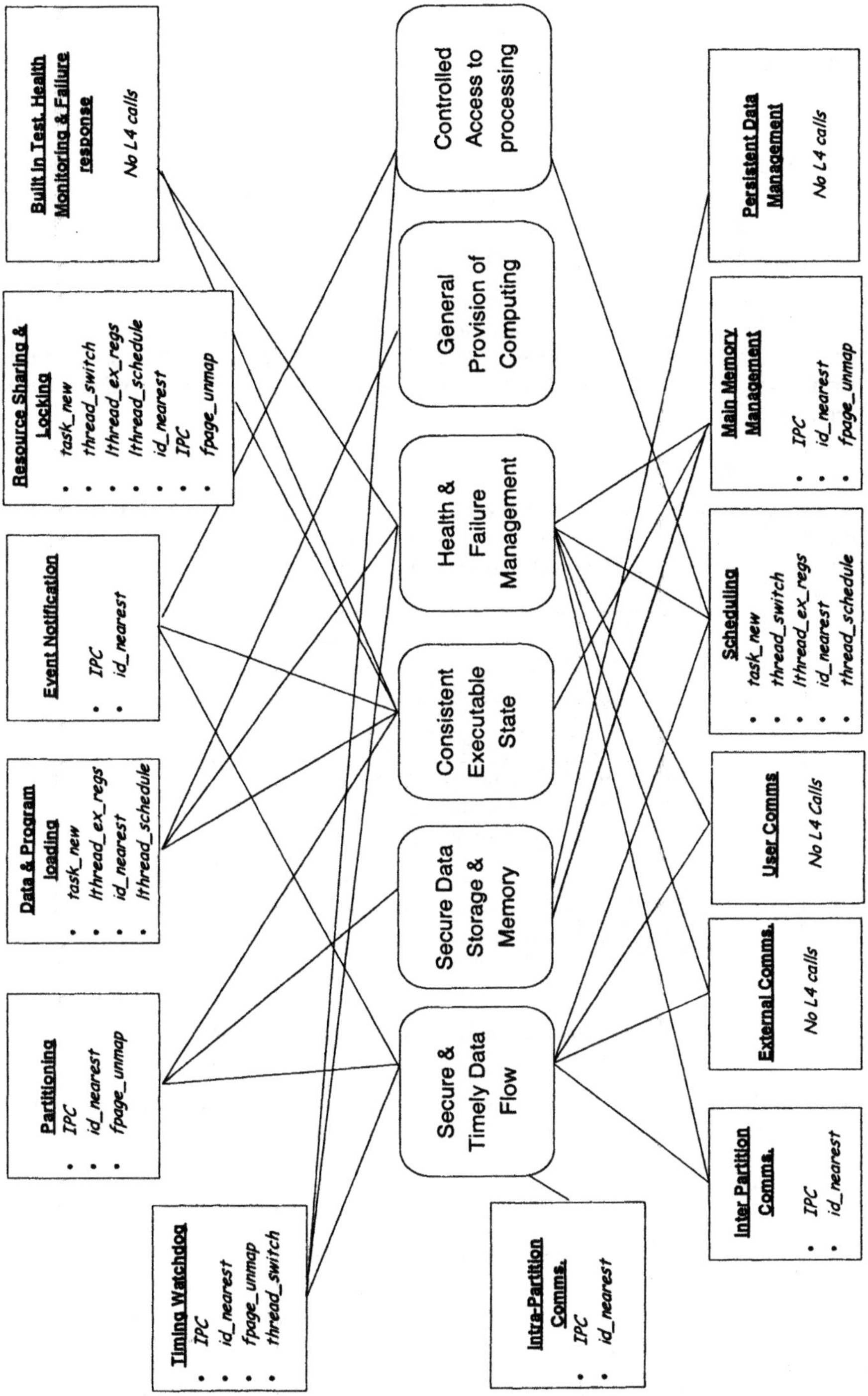

Figure 3: Framework shown populated for L4 microkernel

In addition, we found little or no evidence of code design, i.e. there is limited software design description above the code level as would be required to support AEL 3. This is not to suggest the L4 system is unsuitable for use in an SRS, only that evidence available to us at the time of the assessment would prevent its use if an AEL of 3 or higher were required.

There is a formal verification project being undertaken for L4 (Tews 2001), suggesting that additional evidence may be available to support higher AEL assignments. However, SW01 requires an independent body undertake this assessment activity for AEL 4 and 5 – reinforcing the need for a general assessment framework.

The fact that the assessed AELs are the same for each system call is not especially remarkable as L4 is small and a small team has been responsible for development. The degree of variation in the implementation and testing of each call might be much greater for a larger OS where independent teams may be responsible for implementing different services.

Assigning AELs to services

The AEL for each service is determined by the taking the lowest AEL of the system calls it depends upon. Where no L4 system call supports a service, a '–' is used to show that there is no service. The results are shown in Table 2.

Service	AEL
Partitioning	2
Data and Program loading	2
Event Notification	2
Resource sharing and locking	2
BIT, HM and Failure response	-
Persistent Data Management	-
Main memory management	2
Scheduling	2
User Comms	-
External Comms	-
Inter Partition Comms	2
Intra Partition Comms	2

Table 2: L4 service level AEL assignments

Since all the system calls where assigned an AEL 2, any service supported by L4 should also achieve AEL 2, unless specific evidence emerges that the interaction of system calls supporting a service introduce faults. Our review of evidence for L4 services didn't reveal any such faults.

Derive an AEL for each OS function

As with the services, functions are assigned the lowest AEL of the services it depends upon, unless test evidence at the function level reveals additional faults due to services interacting. The results are shown in Table 3.

Our assessment did not pick up any faults due to services combining to form a function, mainly because the system calls work in isolation and there are simple protocols for message passing and shared memory. Some functions were assigned no AEL due to the lack of L4 support.

Function	AEL	Comments
Secure and timely data flow	-	Single cause: no external or user communications routines
Controlled Access to processing	2	
Secure data storage and memory	2	
Consistent Execution State	-	Single Cause: Lack of BIT and health management routines
Health and Failure Management	-	Many causes: this function would depend on many services – see Figure 3
General Provision of Computing	2	

Table 3: L4 function level AEL assignments

Defining the L4 Taxonomy of Services

Using the populated framework, each function was broken down into services, and each service associated with a number of L4 system calls that directly support it. This provides a taxonomy of services in terms of implemented system calls represented as tables similar that shown for a single function in Table 4.

SUPPORT FOR "Secure Data storage and memory"	**AEL 2**	**References to Evidence**
Service : **Main Memory Management**		See Part 2
API Routine :IPC	AEL 2	
API Routine : id_nearest	AEL 2	
API Routine : Fpage_unmap	AEL 2	
Service : **Partitioning**		
API_Routine : IPC	AEL 2	
API Routine : id_nearest	AEL 2	
API_Routine : fpage_unmap	AEL 2	
Service : **Persistent Data Storage**		
No support	-	

Table 4: Assessment of a function within L4

This shows that a system call is assessed for each and every service it supports. In practice, the AEL of a service will be determined by the rigour and quality of the evidence available for that service. Therefore, in the table, AELs will be copied across the different services supported by that system call. Yet, presenting the results in this way makes explicit the different situations in which the service will be called upon, and allows a decision to be make about the AEL required in each case.

Suitably for specific SRS

On the basis of these results we formed an opinion that L4 was unlikely to be appropriate for an SRS requiring a kernel with an AEL greater than 1. An AEL of 2 could be achieved by either:

- Addition of built in test and some user/external communications facilities with supporting evidence of sufficient rigour to achieve AEL 2.
- Construction of arguments demonstrating the SRS does not rely on any function of L4 assigned no AEL, i.e. not provided.

AEL 3 is unlikely to be attainable without access to the detailed design information. AELs of 4 or 5 could be supported by the ongoing verification effort (Tews 2001), but the assessment would need to be carried out by an independent body.

5. Comments and Recommendations

Assessment framework

Experience during the L4 study showed a framework to be an important tool for demonstrating compliance to SW01, as evidence could only be interpreted meaningfully in the context of the internal structure of the OS. Using this approach, it was possible to determine exactly where the limitations of L4 were in terms of the general services L4 provides. It was then possible to specify conditions under which a higher AEL could be supported by L4.

The assessment framework provides structured way of reasoning about evidence and organising it to allow more direct support for the concepts of SW01 –see Figure 4. By organising the evidence around a template of OS functions and services, arguments can be constructed that are more compelling as the structure of the OS is made explicit, and the implications of changes are more easily assessed. Without such a framework, much of the argument about OS structure is left implicit across the broad array of evidence available. The framework also provides a uniformity of approach that makes use and development of that OS more systematic.

In terms of the original rationale for the framework outlined in section 2, Table 5 below illustrates the progress made in demonstrating the underlying rationale.

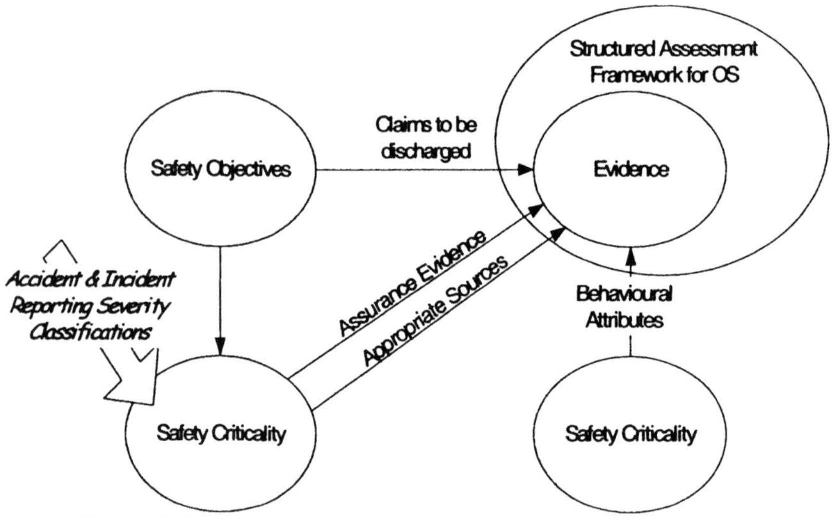

Figure 4: Suggested improvement to the assessment approach

Rationale	L4	Comment
Focusing on services	Y	An AEL was assigned to each service in L4 and implications identified.
Focus on evidence during certification	Y	Evidence has been used to drive the assessment and determine current suitability of L4
Scalability	N	OS and SRS assessments separated, however L4 is not a large OS so no firm conclusions can be drawn about scalability.
Commercial Issues	N	Specific requirements for testing processes and weaknesses identified, however L4 is not a commercial product.

Table 5: Summary of findings

Further work

Table 5 shows the need to exercise the framework using more substantial, commercial OS products. We believe this additional work would be worthwhile as it would improve both the maturity of OS assessment generally and the understanding of SW01 within the ATS domain. There are several research challenges that we anticipate would need to be resolved during this further work:

Relating AEL to safety requirements

There is little guidance available on how safety requirements of SRS applications can be used to derive a required AEL for a given OS function. Appendix A of SW01 provides some initial discussion; however further work will be required. In particular, it would be important to understand the connection between safety arguments of OS functions (derived from the overall system safety requirements) and the AEL assigned to those functions through this assessment process.

Assigning AELs

Whilst the guidance in SW01 is helpful, applying AELs is unlikely to be a straightforward process and involves a degree of judgement. The definition of AELs and the process of assigning them will require a more rigourous definition if the approach is to be used on a wide variety of projects. Further work would define more clearly the process of assigning an AEL.

Commercial Off The Shelf (COTS) OS

Many OS systems will come from a commercial vendor serving a large general market. The framework helps to provide a template against which to assess such systems, however good internal design documentation and an accurate mapping of dependencies between functions and services is required to do the analysis well. If the assessment is made purely on the basis of "black box" evidence and a template assessment framework, there may be a risk that assumptions about OS structure are made that cannot be validated. The framework is an improvement to help structure discussion with suppliers, however commercial barriers will also need to be addressed.

6. Conclusions

Frameworks such as the one suggested in this paper help to improve our ability to assess OS based on evidence available about the underlying services they offer to an SRS. This is particularly useful when applying standards such as SW01 that require we justify the use of OS systems by reference to direct evidence, rather than making weaker arguments about development processes, time in service etc.

The assessment framework provides a way to structure evidence to allow arguments to be made about specific functions, rather than by direct appeal to evidence. The basis of our argument is that functions are less volatile than evidence, and more meaningful in the context of supporting a specific SRS application. The assessment framework depends upon an accurate definition of the OS architecture and by making this explicit, the strengths and weaknesses of a specific OS platform can be assessed. In short, the use of such a framework will help us build stronger, more compelling arguments.

7. Acknowledgements

Thanks to the Civil Aviation Authority Safety Regulation Group for investing in and supporting this work under contract 812.

8. References

(Au 1999) Au A. Heiser G. (1999) "L4 User Manual Version 1.14", Report Number UNSW-CSE-TR-9801. University of New South Wales, Australia

(Conmy 2001) Conmy P and J. McDermid (2001). "High Level Failure Analysis for Integrated Modular Avionics". 6th Australian Workshop on Industrial Experience with Safety Critical Systems and Software, Brisbane, Australia

(CAA 2003) Civil Aviation Authority (2003) CAP 670, "Air Traffic Services Safety Requirements", CAA Safety Regulations Group,pp B3/1- B3/52, http://www.caa.co.uk/docs/33/CAP670.PDF

(Pierce 2003) Pierce R.H., (2003), Nicholson M, Faulkner A.G., "Assessing Operating Systems for Use in Safety Related Systems", International Systems Safety Conference Ottawa 2003.

(Tews 2001) Tews H. Hartig J. Hohmuth M. (2001) "V-Fiasco – Towards a provably correct Micro-kernel", Report Numer TUD-F101-1, Technische Universitat Dresden, Germany

SAFETY ARGUMENT AND THE LAW

The Changing Face of UK Safety Legislation

Julian Hubbard
Independent Safety Consultant

Abstract

This paper examines some of the changes that are currently under review in Europe and which will shortly enter into United Kingdom (UK) legislation. Whilst all European member states will be similarly changing their legislation, this paper focuses on the UK perspective. The General Product Safety Directive demonstrates how consumer safety legislation is changing and the Low Voltage Directive, applied to many products in our field is undergoing radical change. Both are soon to incorporate many of the concepts that were once only found within the domain of safety critical systems management. By using input from the technical committees the reasoning behind many of the new clauses is probed.

1 Introduction

The main areas of UK legislation that impinge on the product safety world are:

- Contract law.
- Tort of Negligence.
- Product liability including strict liability.
- Dedicated legislation, usually for industries which are deemed high risk (e.g. nuclear, railways, marine, electricity supply, etc)
- Occupational health and safety (OHS) legislation (Particularly where products are destined for the workplace but not exclusively so).
    - If the product is going into a UK workplace then your customer will be bound by OHS legislation and will look to you to comply.
    - Installers, maintainers, cleaners, storage and transportation people are also likely to be bound by OHS legislation in relation to your product, hence you must consider the legislation in respect of their activity.
    - UK legislation includes product design clauses that place direct requirements on designers and manufacturers.
    - During the manufacturing phase, aspects of the product design may impinge on its safe manufacture, again OHS legislation must be considered.
- European harmonised legislation (including CE marking) is predominantly aimed at the free movement of goods and trade, but covers the harmonisation of 'Essential Health and Safety Requirements' which could otherwise be a barrier to trade.

Mapped on top of the legislation are technical standards, best practice guides and codes of practice, the appropriate selection of which or application, are important factors in the conformity assessment process and hence have legal implications. All play a major part in defining the products, services and processes businesses manage and supply today. For organisations resident in the UK, without a sound understanding of the legislation, in the UK, the country of manufacture and in target markets, businesses are open to substantial claims if products fail or cause harm. Penalties include product bans, withdrawals, recalls, substantial fines, both on the Company and individuals and even imprisonment of directors and officers.

This is a highly complex area requiring not just an in depth knowledge of the law but also a sound Engineering background in its increasing technical content, together with a thorough understanding of the growing safety management discipline. Apart from specific customer safety requirements included in contract, the customer will assume that the supplier is conversant with the relevant legislation and has taken all necessary steps to comply. Yet how many major companies have this level of expertise in-house or regularly call in lawyers and consultants, let alone small to medium sized enterprises? Much of the information in this paper is drawn from my many contacts in the world of technical committees without who the legislation would be in a much weaker state. Of significant concern is the growing trend not to release these safety experts to carry out this work or even to dispense with their services altogether from their organisations. My greatest worry is that companies will realise this too late when they are facing huge fines, closure, adverse media coverage, key directors having to resign, disgraced products and above all having to face moral questions if people are injured or killed.

There is significant change occurring in the area of European harmonised legislation all of which is transposed into UK legislation. This paper examines the background to two pieces of harmonised European legislation from the perspective of the technical committees and trade organisations where a significant amount of the revision work is done. Due to the fact that revisions and the provision of guidance information are still ongoing it is important to obtain the definitive UK Statutory Instruments when they are available and to follow the existing legislation in the mean time. The paper is only able to highlight a small number of the key issues and does not undertake an exhaustive review of all the changes. None of this review should be taken to imply any legal guidance.

2 The Low Voltage Directive

In the spring of 2001 the formal review of the Low Voltage Directive (LVD) commenced. The initial UK view was that the directive had been in place since 1975 and had largely served the industry well. To embark on a programme of change would potentially open the gates to a long review which in turn could ultimately weaken the directive. The LVD review is continuing and it is hoped that a revised directive will be available in 2004. Certainly the number of changes has been significant, only time will tell if the directive is weakened or strengthened. The following comments reflect some of the key proposals up to draft 3 and since many discussions are ongoing the final directive is subject to change. The review has proceeded well over the past two years but has heralded some proposed major changes. The LVD pre-dated the main body of 'New Approach' (NA) directives and hence has some anomalies compared with other directives. In particular these relate to the removal of the lower voltage limit on the Radio and Telecommunications Terminal Equipment (R&TTE) Directive and the treatment of standards.

The Lower Voltage Limit

Removal of the lower voltage limit (currently 50V ac) would align the directive with the R&TTE directive. Safety aspects addressed by the LVD are not however limited to electrical hazards and the lower voltage limit is not linked to electrical safety criteria such as the Safety Electric Low Voltage (SELV) limit (42.4V peak, 60V dc). Conceivably therefore you could have an electrical product (e.g. powered by a battery or low voltage source) that was outside the scope of the LVD but contained hazards such as high energy levels or hazardous materials, etc. The main question at the time of writing is, does the lower voltage limit get removed or should another arbitrary limit be imposed? There have been many arguments in committee over the limit in order to justify the exclusion of whole groups of products. From an equipment manufacturer's perspective, removing the limit would capture more products within the scope of the LVD of which a significant number would be low risk (in European terminology, 'benign products') and of low monetary value (e.g. singing Christmas cards). All products falling within the scope of the LVD would need some form of conformity assessment process commensurate with the likely risks, together with documentation and marking. All of which could add significant cost if not properly addressed. The current proposal is to add a new Annex (Annex V) to cover a simple procedure based on risk assessment to minimise costs.

Standards

As soon as European standards applicable to the LVD are published (the majority by the European Committee for Electrotechnical Standardisation (CENELEC) or the European Telecommunications Standards Institute (ETSI)) they may be used to demonstrate conformity with the LVD, the title is then published in the European Official Journal (OJ) for information. With all other NA directives, the application of a standard is dependant on its title firstly being published in the OJ. The European Commission has no effective control over the standards bodies. This raised concerns that if a standard was proved to be defective it would be difficult to block its use, certainly within a short time frame because of the difficulties in withdrawing standards. Aligning the LVD with the other NA directives would permit better control through the use of the OJ. The use of international standards (Article 6) and national standards (Article 7) has also come into question. In this latter case, the Member States appear to be split regarding the deletion of Article 7. Major European Standards now include 'A' deviations (deviations for each Member State) which should negate the need for most national standards; however, national standards covering areas such as local wiring regulations will remain and will need to be referenced.

Revised Essential Health and Safety Requirements

The proposed Annex I, 'Essential health and safety requirements' (EHSR) introduces significant numbers of new concepts to the LVD, although not new to the safety world, and adds a large amount of detail to the requirements. This move seems to be going away in part from the original concept of a NA directive aimed at removing detail from the directive and placing it in standards. The issue being that it was easier to update standards to keep them current and hence relevant than to modify European and UK legislation. Designers and manufacturers will therefore need to be familiar with the detailed contents of the directive in addition to standards and guidance material. Some of the key issues are:

- The first point is born out in the Annex I title, the inclusion of '**health**' comes within the scope of the directive. In particular this was prompted by work on electromagnetic fields and radiation.
- The need to carry out '**risk assessments**'. Industry is supportive in that where a 'simple' risk assessment shows that 'benign products' are 'risk free', no further assessment should be required. A risk free product!
- The EHSR are limited to 'intended purposes when properly installed and maintained … which can be reasonably foreseen by the manufacturer for its anticipated lifetime'. This raises the issue of how these aspects can be assessed within a legal context.
- Design and construction now includes a list of categories, each of which is further broken down within the directive. The major categories include:
    - Electrical hazards.
    - Protection to be provided in both normal and single fault conditions.
    - Protection against the spread of fire. At the request of the Danish delegate this may be further expanded to include external ignition sources. This appears to relate to concerns also expressed by the USA into candles placed on consumer products. Of significant concern is that any design solution may be at odds with the new Reduction of Hazardous Substances (RoHS) directive which calls for a reduction in halogenated flame retardants. Also, problems with translating 'fire' into French and German languages need to be resolved.
    - Protection against mechanical hazards.
    - Protection against other hazards.
    - '**Functional safety**'. Includes influences arising from the environment, EMC, software, logic design and power supply disturbances. This caused serious concern within several manufacturers of small household products that they might be drawn into to the rigours of applying IEC61508 unnecessarily and would

have to bear the associated costs. Functional safety is not well understood outside the safety critical world, hence an awareness programme coupled with guidance will be necessary.

- Protection against hazards arising from electric, magnetic, and electromagnetic fields, other ionising and non ionising radiation.
- **'Ergonomics'** including handling and moving the product. A confusing term 'Design for All' has crept into the technical committee papers. The term generally implies taking into account disabilities, the needs of the elderly, children, etc. The draft doesn't expand on this.

Marking & Information

Product traceability remains a major issue within the EU to assist the surveillance activities of enforcement authorities. To achieve this, sufficient details must be available on the product, or if that is not practical, on the packaging to allow the authorities to rapidly trace the responsible person within the Community. Conversely, distributors of branded products don't wish to have a manufacturer's name on the product. Ideally, traceability requirements because they are common to all sectoral directives should be dealt with in a horizontal directive. This was deemed to be too complex to implement given that a significant number of directives and transposed legislation would have to be amended. The wording in draft 3 of the directive still contains confusing terms and requires the manufacturer's name to be included even where they are outside the EU.

3 The General Product Safety Directive

Let us turn to the General Product Safety Directive (GPSD) to get an insight into the European Commission's thinking in the consumer safety arena, remembering that many consumer safety concepts make their way into the professional equipment domain. The original GPSD (92/59/EEC) related to new and second hand products supplied to consumers for private use and was transposed into UK legislation in the form of the General Product Safety Regulations (SI 1994 No. 2328). It specifically did not apply to products used solely in the workplace. This European directive is a 'horizontal' directive in that it applies where no applicable sectoral directive exists or where safety objectives contained in the GPSD are missing from a sectoral directive. The GPSD does not have a CE marking requirement as it is not a 'New Approach' directive. It will therefore be important for businesses to determine the relationship between a revised GPSD and sectoral directives.

Features of the existing GPSD include:

Producers must:

- Provide relevant information to allow customers to assess inherent product risks
- Have systems in place to identify product risks
- Have systems in place to mitigate such risks
- Keep distributors informed on safety issues

Distributors must:

- Not supply products which they know or should reasonably have known to be unsafe
- Participate in monitoring activity, passing on safety concerns back to producers

The original GPSD required a review to be carried out after 4 years and although this was delayed, concern was raised over the general lack of enforcement actions by member States based on the above provisions. A review proposal was submitted to the Commission in March 2000 and following a short review the revised GPSD (2001/95/EC) was adopted by the European Parliament in October 2001. This must be transposed into UK law by January 2004.

Recalls

There will be an obligation on producers to recall products where other measures are not sufficient [Article 5.1]. Enforcement bodies may order a product recall as a last resort [Article 8(f)(ii)]. It was felt that some enforcement bodies would be reluctant to enforce a recall for fear of financial penalties arising from a successful appeal unless protective measures were in place. Additionally, the GPSD requires that when enforcement action is taken they should act "taking account of the precautionary principle" (see below). Whilst UK industry opposed in principle mandatory recalls, claiming that voluntary recalls worked well under the existing system, consumer group pressure pushed for a formal system and won the day. As soon as you move to a legally enforceable recall requirement many issues come to the fore including:

- Who determines whether a recall is necessary and what liability attaches to that person?
- What processes should be followed? How are these validated? What if the process subsequently is shown to be defective?
- How do you set and obtain legal agreement on acceptable rates of return and duration?
- How do you close down a recall?
- If a recall is imposed by an enforcement authority, how do you lodge an appeal?
- What happens if Member States have differing views on recalls?
- If a recall is mandated and a firm becomes insolvent, who pays?

Both the UK Department of Trade and Industry (DTI) and British Retail Consortium (BRC) have produced good practice guides on consumer product recalls in conjunction with UK businesses and trade associations. These could be built on at an EU level to produce harmonised Commission best practice guidance on managing a recall and enforcement.

Notification and Dangerous Products

Producers and distributors must notify dangerous products to the enforcement bodies [Article 5.3 and Annex I]. Although rules still have to be drawn up, concern was raised that a Europe wide body to disseminate such information around the national bodies hadn't been included, businesses would therefore have to notify each individual enforcement body where the product was marketed or likely to be deployed.

Further concerns have been expressed over the reporting threshold and associated penalties. Minor problems could be brought to the attention of a producer, which after investigation prove to be more serious and products are then subsequently reported to the authorities. If reasonable uniform threshold guidelines are not set across the EU then that producer could be liable for not reporting the problem earlier. The US model was a major influence in preparing this article. In the USA, potentially dangerous consumer products must be reported immediately (within 24 hours) to the Consumer Products Safety Commission (CPSC) as soon as the business is aware of a problem. This may occur before the business has completed its own safety investigations. Failure to notify the CPSC can result in severe penalties even if the product is vindicated following an investigation. The American system has a major benefit over the European system in that it is centralised. Not only is reporting into a single body, but enforcement action originates from that body. In Europe, each Member State is responsible for its own enforcement actions and producers may have to deal with several bodies for the same enforcement requirement.

Dangerous products which are subject to an 'Emergency Decision' will be banned and may not be exported [Article 13.3]. The aim of this clause was to prevent dumping on countries outside the European Union (EU) that didn't have such stringent safety legislation limiting the importation of potentially unsafe products.

Other key changes embodied by the new Directive:

- Products intended for use by professionals which may subsequently be supplied to consumers ('migrating' products) come within scope [Article 2(a)].
- Product safety assessment will additionally include putting into service, installation and maintenance requirements [Article 2(b)].
- European standards will now have to be established and their titles published in the European Journal [Article (4)]. Products in conformity with those

standards will be deemed to satisfy the requirements of the directive. Concern has been raised that where a sectoral directive also applies standards, standards specifically excluded from that directive could be re-applied by virtue of having to apply the GPSD.

- Distributors must keep and provide documentation to assist in a product recall [Article 5.2]. Questions were raised over the application of the Data Protection Act to the retention of customer details, at the time of writing the jury is still out. No information is given regarding the issue of document/ data retention periods and requests for firm guidelines have been made. Consumer groups lobbied for a 10 year retention period, businesses argued that it should be less than 6 years. They also argue that data retained on a purchasers contact details ages rapidly in the consumer sector as people move between properties, products are prematurely disposed of and ownership changes.
- Whilst services are still excluded, although these may be the subject of separate legislation in the future, products supplied in the course of a service are within scope [Article 2(a)].
- Products which conform to the requirements of European standards which set an appropriate level of safety will be presumed to satisfy the GPSD's general safety requirement.
- New terms entered into the legislation for the first time including "serious risk", "recall" and "withdrawal". Calls have been made for these to be fully defined in guidance information.

A strong emphasis is placed on supplementary guidelines. However, as with other directives, guidelines are non-binding and have no legal standing other than potentially setting best practice.

The revised directive goes on to say that Member States must establish or nominate authorities to ensure that the directive is effectively implemented. Again, measures are not stated and in the UK it would imply a significant expansion in resources, predominantly in the Trading Standards organisation. Most other Member States have similar resource issues. It will be interesting to see if the required level of resources is made available in a consistent manner across EU states.

For most businesses that have comprehensive product safety management systems to manage risk, monitor field safety incidents and handle recalls, the formalisation of the requirements set out in the GPSD should not impose significant further burden. Other organisations will need to respond rapidly to implement such a system. The new directive and transposed UK regulations contain a significant amount of detail in relation to product safety management in the consumer sector. It is essential that those Companies working in that sector, both producers and distributors become fully conversant with the detail text and implement appropriate management systems. Organisations will also need to track and understand the massive amount of guidance material that will be released over the forthcoming year and monitor the activities of the enforcement bodies for potential case law. Legal firms and consultants are already staking their claims to a slice of this information provision.

4 The Precautionary Principle

The GPSD makes reference to 'The Precautionary Principle'. This principle is being increasingly referenced in matters relating to safety, particularly long term health (this includes human, animal, plants and damage to the environment) and is finding widespread application and influence, so what is the principle and what were its origins?

Firstly, there have been several versions of the principle. Principle 15 at the United Nations Conference on environment and Development in Rio de Janeiro in 1992 advocated the widespread application of the precautionary principle. In 2000 the European Commission issued a communication [COM(2000)1] aimed at clarifying the principle and its application within the EU.

The principle requires that where complete scientific understanding, consensus or certainty is still to be established and reasonable suspicion of harm has been noted arising from some (possibly unknown) agent and we do not fully understand the causal link between the agent and the harmful effect then we should apply cautionary controls. The major concern is where products or processes are launched that could not be subsequently controlled or recalled and could give rise to long term health issues that could not be reversed. Basically, before a product is introduced we must be able to demonstrate that the safety risks are small and outweighed by any benefits. Until such time as the evidence can be presented we should not expand our use of that technology and preferably reduce dependence on it. The burden of proof is therefore shifted to the innovator or designer to prove beyond reasonable doubt that a product or process is safe. The principle does not say that absolute proof is necessary, as we know all scientific and technical areas carry a finite residual risk.

One of the problems is that the principle is open to interpretation and can be misused to justify the imposition of much stricter controls than would be justified through scientific understanding and risk assessment.

The principle has its roots back in the environmental sector with links to food, genetic engineering, global warming, pollution, etc. In the UK the principle moved to centre stage following the BSE crisis and release of the 'Stewart Report' into mobile phones in May 2000. The report followed an enquiry by the Independent Expert Group on Mobile Phones (IEGMP), chaired by Sir William Stewart into the safety of mobile phones and base stations and potential links with electromagnetic fields (EMF's) in causing problems with health.

Increasing application of the precautionary principle may see a growth in sound risk management practices, particularly where a regulator is involved. Processes already adopted by responsible organisations may become formalised to comply with regulatory requirements. For example:

- Research to reduce uncertainty
- Risk assessment
- Risk reduction
- Hazard avoidance/ substitution
- Communication of residual risks

With these may come greater burdens on industry:

- Stronger enforcement
- Larger penalties
- Potential bans on certain technologies, products or activities
- Requirements to provide information, publicity, guidance or information
- Health monitoring

One of the cornerstones of the principle is the need to make a decision based on sound scientific evidence rather than allow products to drift into the market.

It is important that clauses are built into legislation permitting the development and release of products that may contain elevated risks where these replace existing products having greater risk or the risks outweigh the benefits to society.

5 Conclusion

What about the future? With the expansion of the EU in 2004 and the huge quantity of work that that will entail it has been rumoured that some of the committee work may become sterile and put on hold until the hiatus settles down. Given that we have embarked on and are well through a review of the LVD my wish, and I believe that of others in the committee world is that agreement is reached quickly on the proposed revisions and that the transposition into national legislation is not delayed.

Can the directive review process survive an expansion of the EU in its current form? Consensus between the 15 Member States is difficult to achieve in a reasonable time frame. I believe that we will need to seek out different ways of working when the expansion occurs.

If sensible revisions are to be undertaken to European legislation in the future then those competent in product safety management must be released to sit on technical working parties.

Companies must ensure that they identify and understand the impact of all relevant legislation on design programmes if disaster is to be averted.

6 References to legislation

The Low Voltage Directive (73/23/EEC as amended by 93/68/EEC)
Implemented in the UK by:
The Electrical Equipment Safety Regulations 1994 SI 1994/3260
Guidance – Product Standards (Implementing the Low Voltage Directive)
Guidance Notes on UK Regulations – DTI July 1995 (URN 95/626)

The General Product Safety (GPS) Directive (92/59/EEC)
Implemented in the UK by:
The General Product Safety Regulations 1994 SI 1994/2328
Directive 92/59/EEC has now been revoked and is replaced by 2001/95/EC
http://europa.eu.int/smartapi/cgi/sga_doc?smartapi!celexapi!prod!CELEXnumdoc&lg=en&numdoc=32001L0095&model=guichett

Communication from the Commission on the precautionary principle COM(2000)1 2/2/2000
http://europa.eu.int/comm/environment/docum/20001_en.htm

Inter-Departmental Liaison Group on Risk Assessment (ILGRA)
The Precautionary Principle: policy and Application
http://www.hse.gov.uk/aboutus/meetings/ilgra/pppa.htm

Rio declaration on environment and development, made at UNCED 1992, ISBN 9 21 100509 4
Principle 15
http://www.unep.org/Documents/Default.asp?DocumentID=78&ArticleID=1163

Turning Up the HEAT on Safety Case Construction

Paul Chinneck
Safety & Airworthiness Department
Westland Helicopters, Yeovil, BA20 2YB, UK
chinnecp@whl.co.uk

David Pumfrey, Tim Kelly
Department of Computer Science
University of York, York, YO10 5DD, UK
david.pumfrey|tim.kelly@cs.york.ac.uk

Abstract

The HEAT/ACT project consists of replacing the conventional mechanical flight control system of a helicopter with a fly-by-wire system. With such a project, the safety concerns are obvious, and therefore the development of a thorough and convincing Safety Case is paramount. Goal Structuring Notation was chosen as the method for this, on its perceived merits of ease of construction and clarity of review. This paper outlines the work conducted, and appraises these perceived merits against experience during and following the construction of the Preliminary Safety Case.

1 Background to the HEAT/ACT project

The HEAT/ACT project consists of replacing the conventional mechanical flight control system on a helicopter with a fly-by-wire system (see Staple & Handcock 2002). It involves extensive re-engineering of the aircraft systems, including:

- removal of two out of three hydraulic systems
- replacement of the main and tail rotor hydraulic actuators with electro-mechanical actuators
- removal of mechanical flying controls.

The major new items include:

- adding another electrical generator
- installing actuator control units
- adding two new fly-by-wire (FBW) flight control computers.

1.1 Safety case approach

With such a project, the safety concerns are obvious, and therefore a convincing and thorough Safety Case is paramount. Def Stan 00-56 (Ministry of Defence 1996) Part 2 "encourages the concept of an evolving Safety Case" in order to "[initiate the Safety Case] at the earliest possible stage…so that hazards are identified and dealt with *while the opportunities for their exclusion exist.*" The HEAT/ACT project recognised the benefit of constructing the safety argument as early as possible, for this precise reason. Any change to the architecture, for safety reasons or otherwise, becomes dramatically more difficult and expensive once designs are frozen and components are in manufacture. The project therefore chose to adopt a phased approach to safety case development, beginning with a Preliminary Safety Case (PSC).

The ultimate intent of the Safety Case was to show, by clear argument, that the HEAT/ACT system, when fitted to a Merlin helicopter, will be acceptably safe for operational use. The PSC was required to contain a complete argument, showing all of the important claims that would have to be made and demonstrated in order to satisfy the safety requirement. Goal Structuring Notation (GSN) (Kelly 1999) was chosen as the method for representing this argument on its perceived merits of ease of construction and clarity of review. Early proposals for evidence to support this argument were also required as part of the PSC development.

Through a process of review with airworthiness authorities, the PSC would provide confidence that the design of the HEAT/ACT system was such that acceptable safety could be demonstrated. It would support the engineering process by providing a structure in which to document safety activities and processes until the system has been fully designed and final safety analysis has been completed. Before the system hardware is produced and commissioned, an Interim Safety Case will be required. This will use the PSC as a basis, refining and adapting it to suit the final system design, and adding more complete evidence to support the argument.

The scope of the Safety Case is to include an assessment of the impact of the modification on the safety of the whole aircraft. It is of course of no benefit to show that the new system is acceptably safe without also considering the effects it has on the platform into which it is integrated.

1.2 Timeline

Guidance (Kelly et. al. 2003) shows that the pre-requisites for preparation of a PSC are:

1. The initial Safety Programme Plan (SPP) for the project (though note that this will develop as the PSC is constructed, as evidence requirements will generate new tasks that will have to be incorporated in the SPP)
2. A Preliminary Hazard Identification (PHI), together with initial hazard analysis and Risk Estimation
3. Identification of key safety requirements

This guidance also states that the PSC should be constructed before:

1. Detailed specifications are produced
2. Detailed design and implementation work (and any related analysis and testing) has commenced

3. System Safety Analysis (following on from Preliminary Hazard Analysis) is conducted

A stated requirement of the HEAT/ACT project SPP was that the PSC should be prepared to support Preliminary and Critical Design Reviews (PDR and CDR). Producing the PSC would also ensure that the Interim Safety Case could be prepared in time to support first flight.

The timing of the request for HEAT/ACT PSC production met both the above guidance and the project requirements. In the programme of a typical project, this should have allowed many months in which to draft, develop and finalise the PSC. However, the HEAT/ACT project has very challenging timescales, giving no more than two months for construction and issue of the PSC.

1.3 PSC development and presentation

The HEAT/ACT PSC was largely written by a single person. Throughout the development process, however, the expertise of others was called upon to review the argument. Several individuals read and commented on the document as work progressed, and there were also more formal meetings in which the primary author worked through the structure of the argument with a group.

The GSN diagrams form a key part of the delivered PSC document. The PSC starts with preliminary material, including an outline description of the HEAT/ACT system, and an introduction to GSN notation to help readers who are not familiar with its use. The majority of the document then presents a small section of the argument in GSN on each page, followed by textual discussion. The text explains the intent of the argument fragment, with justification where necessary, and describes how it is intended that the fragment will be developed (including an outline of evidence requirements) in the Interim and subsequent phases of Safety Case development.

2 Outline of the experience

2.1 Starting out

The definition of the top goals of the GSN structure was relatively easy, as much guidance exists for starting safety arguments. The system and its integration into the aircraft would be treated as separate concerns. For the system argument, the common "product/process" argument approach was taken. In the "product" argument strand, the identified system hazards are addressed, and developed until direct evidence can be presented to show that the risks from these hazards have been adequately controlled. The "process" strand demonstrates how all applicable standards and requirements are satisfied.

However, after quickly sketching out the top few levels of the argument structure, it became increasingly difficult to extend the argument structure into lower levels of sub-goals. A number of different situations were evident. In some cases, it was already obvious what the solution (evidence) nodes at the bottom of the argument should be; the difficulty here was in "reaching" existing evidence, i.e. breaking down high-level goals to sub-goals at a level where it could be shown that the evidence directly and sufficiently satisfied specific goals. In other parts of the structure, the problem was effectively lack of inspiration; some of the goals clearly required significant decomposition, but no obvious strategy could be seen. A third problem was how to resolve obviously related goals in different branches of the argument structure.

2.2 Developing the argument

At this point in the development of the PSC, advice was sought, and both safety case patterns (Kelly & McDermid 1997) and a number of existing safety cases were considered as potential sources of inspiration and guidance.

2.2.1 *The patterns approach*

Safety case patterns represent sections of argument to satisfy specific goals. They require "instantiation" to suit the context into which they are introduced; that is, they contain elements such as alternatives from which the appropriate selection must be made, and incomplete structures, which must be extended. In applying patterns in the context of the HEAT/ACT PSC it was found that, at the top levels of the safety case structure, they generally yielded good guidance and suggestions for the expansion of specific goals. Specific patterns incorporated into the argument included Functional Decomposition, and the commonly used (but possibly not fully understood) ALARP.

As the project progressed, however, it was found that patterns were of more limited use in developing the argument to a sufficient degree to complete any part of the GSN. This may be because patterns are, by their very nature, "skeleton" structures, which require tailoring to be properly used. Without experience of identifying appropriate relationships between the "reality" of the system or process that is to be represented, and the library of available patterns, it can be difficult and time-consuming to make appropriate choices. Also, considerable expertise is required in deciding when a pattern has been fully instantiated, and a branch of the argument can really be considered complete.

What proved more useful at this low level was actual safety case material. This yielded the ability to examine how other authors had "solved the problems" of instantiating patterns, and also suggested "micro-patterns" – often a very small number of GSN elements, or even just well-chosen wording of a goal or strategy – which provided the inspiration to assist development of the argument to the very lowest level.

2.2.2 *Borrowing and modifying*

In investigating how to complete the HEAT/ACT PSC argument structure, the authors examined a number of published safety cases that have used GSN. Two specific safety cases were found to be extremely helpful; the European Reduced Vertical Separation Minimum Pre-Implementation Safety Case (EUR-RVSM PISC) (Eurocontrol 2001) and the safety argument for integration of new technology on to older platforms proposed in Kenneth Graham's MSc project report (Graham 2002). Examples of ideas adopted from these documents included:

EUR-RVSM PISC:
- Proving that any requirements were firstly complete and correct, and then showing that those correct requirements were satisfied.
- Variants on the "direct evidence supported by backing evidence" pattern. (A similar pattern is also used extensively in the UK Civil Aviation Authority's standard for Software Safety Assurance (Civil Aviation Authority 2001).)
- Stating Safety Objectives at the beginning of the Safety Case, proving they are addressed within the safety argument, then showing compliance in a summary at the end.

Kenneth Graham's MSc project report:
- Integrating a system under fault-free conditions, then assessing the fault tolerance or failure management following installation.
- The key assertion that a system can only be safe after modification provided that the parts of the system not affected by the modification were acceptably safe in the original state.

All of these ideas provided important "building blocks" for the safety argument. The authors also found that reading and review of these and other safety cases were significant in building confidence in our own ability to recognise – and subsequently produce – well-structured and robust arguments.

3 Successes

3.1 Speed of construction

The request for the PSC was made in early April 2003. Development of the argument by the primary author began in mid-April, but quickly reached an impasse. In mid-May other expertise was drawn in to review the work so far, and the approach described above using patterns and drawing on other safety cases was adopted. From this point, the rate of progress dramatically increased. Within a further two weeks, the document had grown from just 8 very basic GSN diagrams to 26 much more comprehensive ones, for presentation at project PDR. A further three weeks' work saw the document ready for signatory approval. From this state to the final issued version, only very minor changes were required, despite requiring four different people to check and sign the PSC as approved.

3.2 Development and discussion

As the document progressed, it became clear that using GSN had some huge advantages over a textual document:

- Many of the revisions required during the development were merely expansions of earlier work, rather than complete re-writes of sections. This is illustrated in the sequence of Figures 1 to 3, where three stages in the expansion of the Hazard Log goal can be seen. The final expansion of this goal in Figure 3 actually forms a very satisfactory pattern, which could be used in other safety cases where maintenance of a hazard log is an important aspect of safety management.

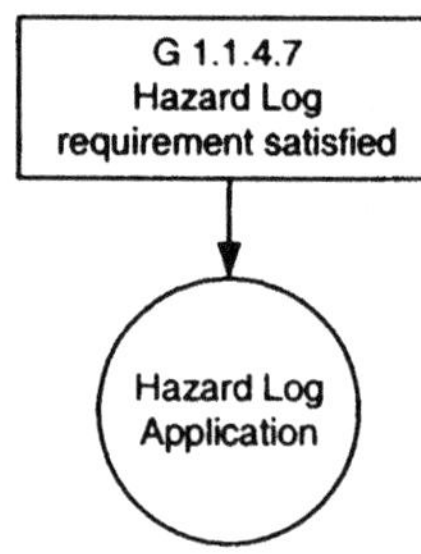

Figure 1 - Stage 1 of evolution of Hazard Log requirement GSN

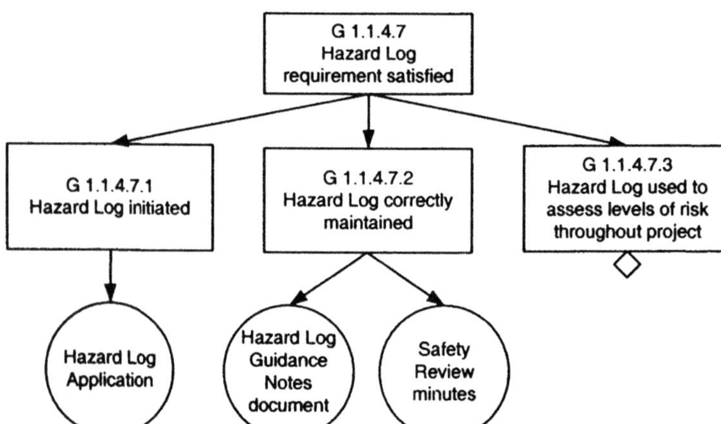

Figure 2 - Stage 2 of evolution of Hazard Log requirement GSN

- It was possible to re-use self-imposed "patterns" from earlier in the document. As a new area came under scrutiny for development, thoughts and methods from earlier on were re-used, dramatically reducing the time required.
- It remained very simple to keep the "big picture" in mind, even when developing the structure to six or seven sub-levels.

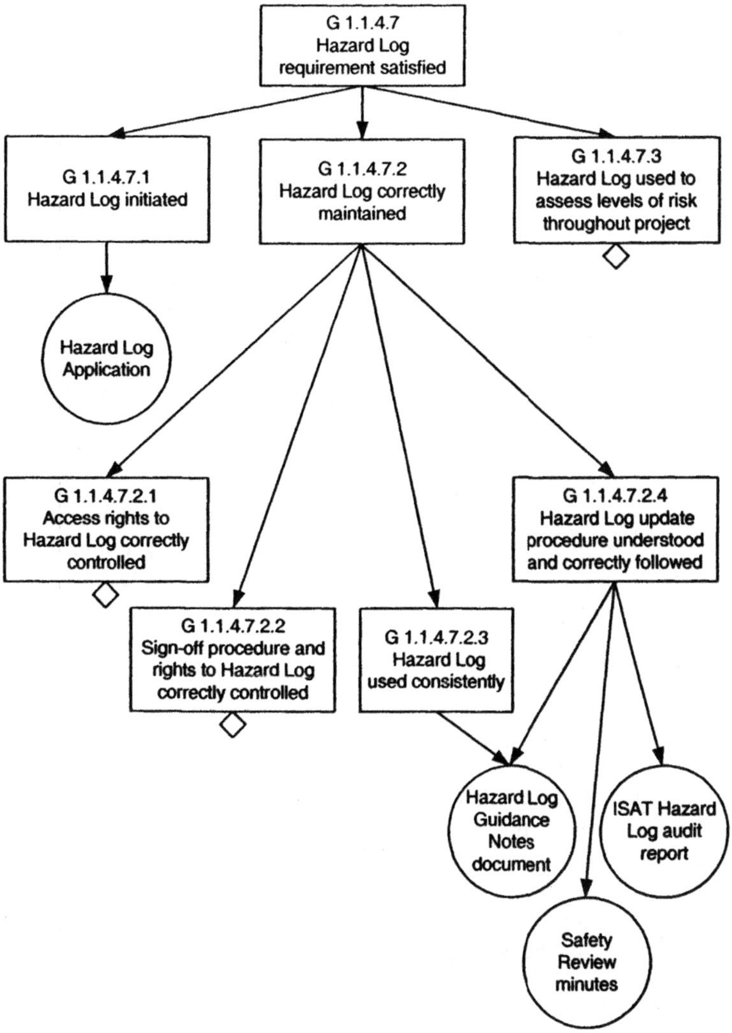

Figure 3 - Stage 3 of evolution of Hazard Log requirement GSN

- It was much less likely that areas of the safety argument would be overlooked. Working with top-level goals before breaking them down into sub-goals helped this enormously.
- Discussion of parts of the document was simplified, as it was easy to use the GSN hierarchy to explain to people the context of the area under discussion.

A further benefit of GSN was realised when it came to asking equipment suppliers for sub-system safety cases. The diagrams showing the breakdown of hazards related to system functions formed an easy starting point for helping each supplier to identify their contributions to the safety case. Various parts of the process argument were also used to show the relationships between different organisations' safety activities.

3.3 Successful reviews

At various stages throughout the creation process, Safety Case reviews were undertaken. Again, the benefits of GSN became clear during these reviews. At one point, the near-complete document (comprising 26 pages of GSN diagrams) was presented to representatives from the MoD at the project PDR. This took just 30 minutes.

The document was also reviewed at the various levels of Project Safety Meetings held. Not only was it relatively quick to review the entire Safety Case, but it was easy to see what had changed or been added since the last review. Conducting these on-line reviews on a text-based document would have been virtually impossible. At a practical level, a single GSN diagram could easily be "lifted" from the document and made into a projection slide as the basis for discussion.

Following issue of the document, the primary author received a number of complements relaying how simple the document was to read and understand. These comments came from both technical and non-technical staff, proving the benefit of GSN in representing the argument in a clear and unambiguous fashion.

4 Problems and suggestions

4.1 Giving the right impression

The regular reviews of the GSNs in the development of the project drew attention to a number of issues of presentation. One of the most important of these was the relationship between "reading order", and the impression of the overall argument being made. At a fairly early stage in developing part of the argument, a relevant pattern with the structure shown in Figure 4 was identified in one of the sources, and instantiated into the safety case. When this part of the safety case was discussed in review, one of the participants objected, as he had perceived the argument to be primarily about safe behaviour following a single failure, whereas he felt that safe behaviour in normal circumstances was more important.

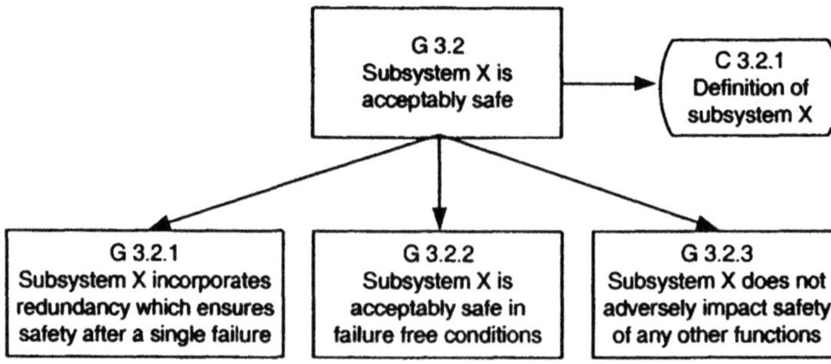

Figure 4 - GSN showing "perverse" ordering of goals

In GSN, layout of items on the page does not imply any order of importance or precedence. Thus, the sub-goals in the pattern above should all be considered to have equal importance in the argument. However, it is worth bearing in mind that the reader's natural tendency is to consider the sub-goals in "normal" reading order, i.e. left to right across the page. Particularly in the mind of readers who are not familiar with GSN concepts, the order in which sub-goals are encountered will inevitably form an impression about their relative importance. In this case, although the components of the argument are all correct and acceptable, the goal relating to safe behaviour following a single failure is encountered before the goal relating to safe behaviour in normal circumstances, giving the incorrect impression that the former is a more important component of the argument.

Pragmatically, therefore, when laying out arguments in GSN form, it is necessary to think about the way in which a reader will approach the document, and perhaps adjust the presentation accordingly. If it really is the case that some sub-goals are more important, then consideration should be given to using a strategy element to identify this ordering. For example, in building the HEAT/ACT safety case, there were a large number of instances where a generic pattern of satisfying a goal through a combination of direct evidence supported by "backing" evidence was used. In these instances, it could be seen that the direct evidence was "spinal", i.e. the safety case would collapse without it, whereas the backing evidence was of lesser importance; inability to provide one of these items of evidence might weaken the argument, but not fundamentally destroy it. In these cases, we considered it might be beneficial to incorporate an explicit strategy element into the GSN, as shown in Figure 5. A similar structuring effect could alternatively be achieved through the use of appropriately worded sub-goals to explicitly differentiate direct and supporting evidence.

In discussing the presentation / reading order of GSN elements, it is also worth considering the way in which the GSN will be "flattened" to transform it into the structure of a full textual safety case. Here, even more than in the diagrammatic form, the order in which items are presented will form a strong impression with the reader. Careful consideration must therefore be given to the order in which the GSN tree is traversed to construct the contents list for the text. We note that at least one of the commercially available GSN tools (Adelard's ASCE (Adelard 2003)) allows this sequence to be specified.

4.2 Representing process

At a high level, the HEAT/ACT safety case made use of the classic "product / process" argument pattern. However, as the argument was refined and developed, we encountered a number of areas where this split proved problematic to develop completely.

At first sight it would appear that, provided the "product" side of the argument is satisfactory, there should be little difficulty in completing the "process" branch. Intuitively, since following the defined safety process produces the evidence

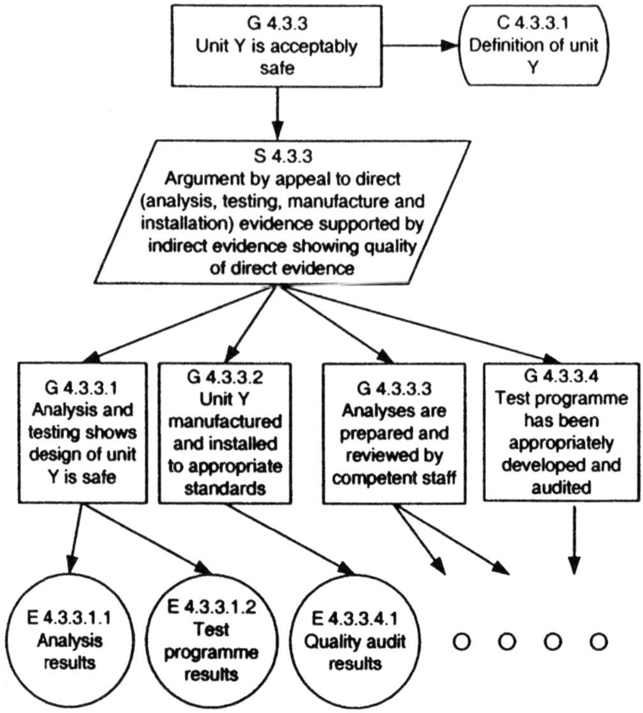

Figure 5 - Using a strategy to highlight the use of direct and backing evidence

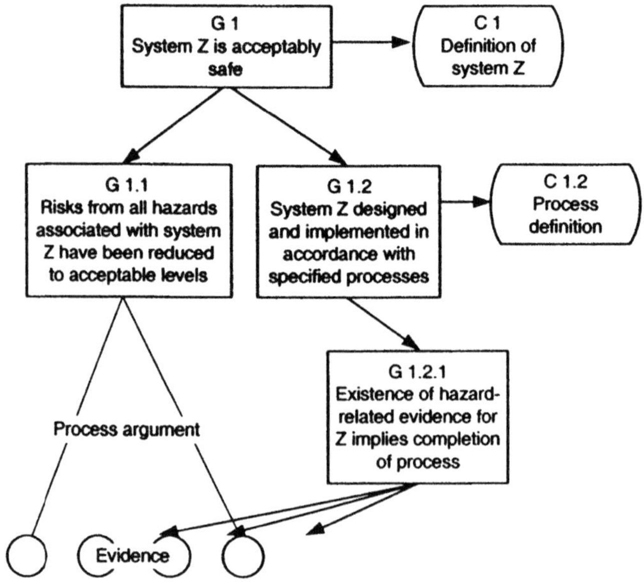

Figure 6 - Naive product / process structure

required to argue product safety, the existence of the product evidence should imply satisfactory completion of the process. In GSN notation, we might expect an argument with the structure shown in Figure 6, where the same evidence is used to satisfy both product and process argument strands.

In practice, however, we found a number of difficulties with this structure. The first is that a safety process is complex, involving many stages. Unless it is very carefully captured and maintained, product-structured evidence may not provide sufficient information to demonstrate that all the steps of the process have been followed. Consider, for example, a "product" safety argument that the risks presented by all identified hazards have been reduced to an acceptable level. The top few levels of such an argument structure are shown in Figure 7. If we also want to demonstrate that a classical safety process has been followed, we may want to complete a structure such as that shown in Figure 8.

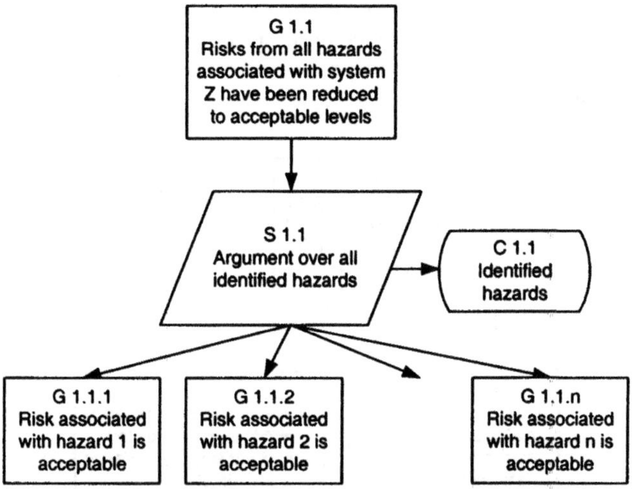

Figure 7 - Hazard-directed argument fragment

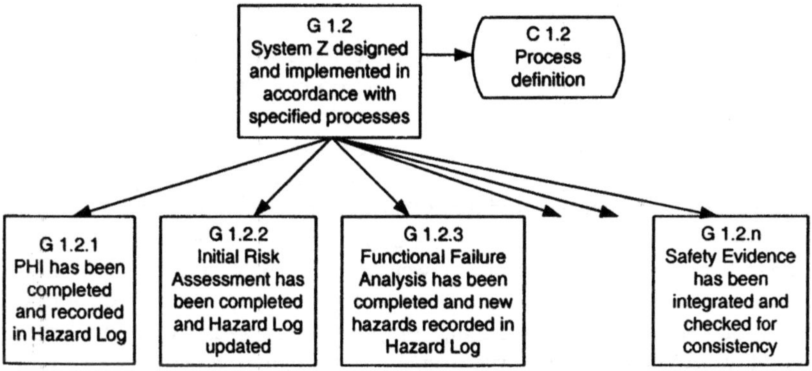

Figure 8 - Process steps argument fragment

The problem in reconciling these two argument approaches is that, as the hazard log is a dynamic document, work done in later parts of the process might update, obscure, or even totally replace, work done at early stages such as PHI. Thus, unless we ensure that a complete change history is available for the hazard log, its final state is unlikely to reflect all of the work that has actually been done. This is a real problem; one of the authors was involved in reviewing a safety case for another project in which exactly this problem was encountered; the safety case claimed that the final hazard log for the project was sufficient evidence that the defined process had been followed; in reality, so many changes had been made during the project that it was impossible to gauge the quality and completeness of the early work.

The benefit of identifying and considering this issue in developing the PSC is that decisions can be made about how to ensure complete process evidence is available. The GSN argument fragment in Figure 9 shows the general approach used in HEAT/ACT, based upon means of compliance matrices. For each part of the process, a choice has to be made whether to ensure complete history is retained in the final state of each item of evidence, or whether "baseline" versions will be retained at key process points.

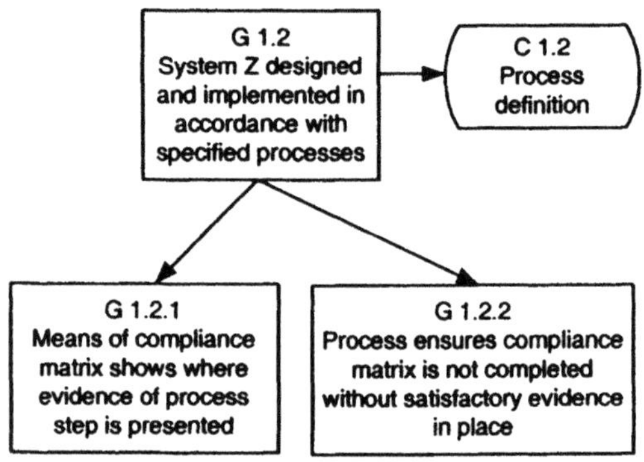

Figure 9 - Argument fragment showing how product evidence is used to support a process argument

Another process-related issue that was encountered in outlining the HEAT/ACT PSC was the occasional temptation to include too much process information in the safety case outline. This was evident, for example, in the first attempt to show how explicit top-level safety requirements would be satisfied. This divided the requirements into two classes – those that could be satisfied directly by compliance with airworthiness directives, and those that were more complex and would require exceptional activities or the involvement of specialists. As the argument structures for these two classes of requirement were developed, it became clear that their evidence requirements were essentially identical. The split had been introduced because of the strong perception that they would be satisfied in fundamentally different ways; whilst this was true from a process management perspective, it was an unnecessary complication in the safety case.

A key element of the "process" strand in the HEAT/ACT PSC is showing compliance with appropriate standards – notably Def Stan 00-56. Attempting to demonstrate this in the GSN presented a number of problems, foremost among which was a direct conflict between easy "read-across" (i.e. presenting a structure that closely reflected relevant sections of the standard), and following a "natural" breakdown (i.e. presenting a structure which reflects the actual organisation of the project). The chosen solution is something of a compromise. Major process elements, such as the key activities shown in Figure 8, which are fundamental to achieving safety and would have been implemented whichever standard was being followed, have been developed in a natural structure. However, where additional arguments and evidence have been included specifically to satisfy particular requirements of the standard (e.g. the elements of the Safety Management System required by chapter 5), these have been presented in a structure (Figure 10) that directly reflects the organization of the standard.

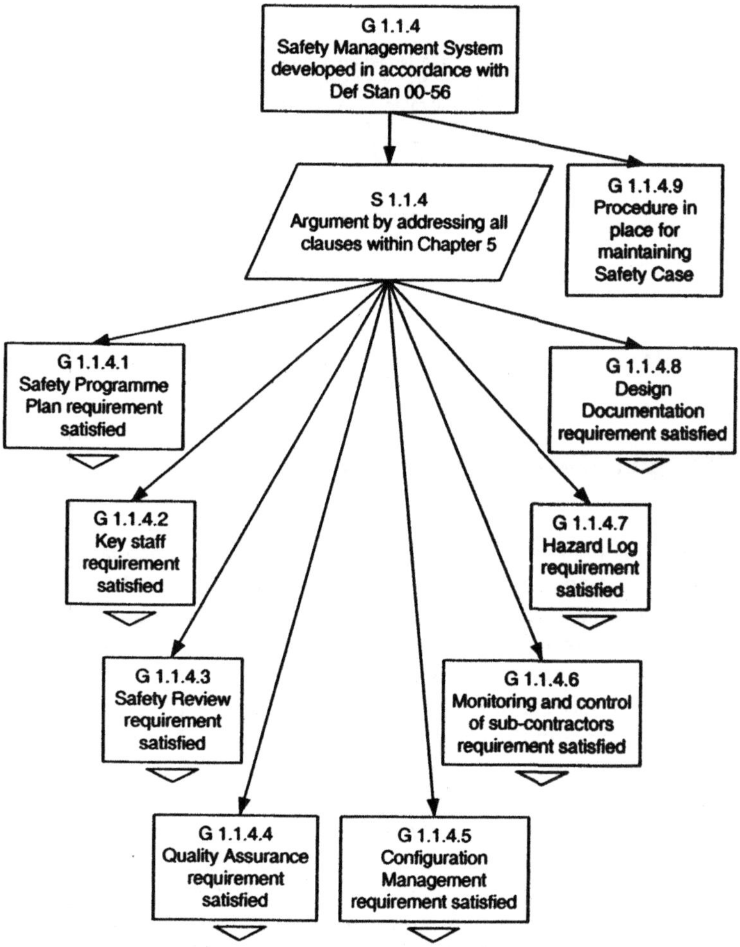

Figure 10 - Satisfaction of SMS requirements from Def Stan 00-56 chapter 5

We discovered that, in addressing the Safety Management System (SMS) requirements of Def Stan 00-56 chapter 5, we needed to create and instantiate another small pattern. The paragraphs in this chapter require various management documents and controls, such as the preparation of a system safety programme plan and the implementation of a system of independent audit. In considering how to show satisfaction of these requirements, we recognised that it would be possible to show that the "letter" of Def Stan 00-56 had been followed merely by pointing to the existence of the relevant items. However, to satisfy the "spirit" of the standard actually requires three separate elements; demonstration that the SMS item has been created, that it has been maintained (i.e. reviewed and updated as the project evolves), and that it has actually been followed. We therefore created the pattern shown in Figure 11 to show satisfaction of each of the paragraphs. An example of the instantiation of this pattern can be seen in the Hazard Log requirement argument in Figure 3.

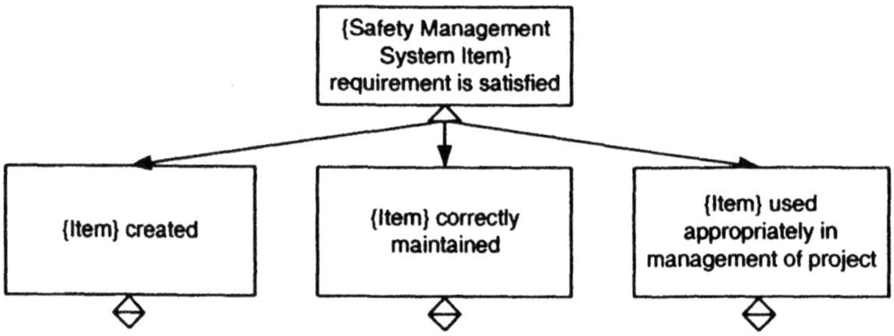

Figure 11 - Pattern to demonstrate satisfaction of SMS Item requirement

4.3 Managing size and complexity

As might be expected, the HEAT/ACT PSC is already a substantial document. The actual size of a safety case cannot be limited – it would be completely unacceptable to exclude information simply on the grounds that "it's already too big" – and elaborate argument structures are unavoidable when examining complex systems. Consideration therefore has to be given to practical means for managing size and complexity. In the HEAT/ACT project, we used two related mechanisms that we believe have been successful in making the safety case simpler to navigate. A particular problem area in the classic hazard-structured GSN argument has always been the point where the argument is split into branches for each individual hazard (as in Figure 7), as this causes an "explosion" in the width of the structure. In the HEAT/ACT PSC, it was recognised that the hazards could be grouped by system function, and this organisation was therefore used as an intermediate step in expanding the argument over all the identified hazards. One of the key benefits we expect to obtain from this structure in making the full safety argument is that there will be significant commonality between the evidence presented for the hazards within a group, as sketched in Figure 12.

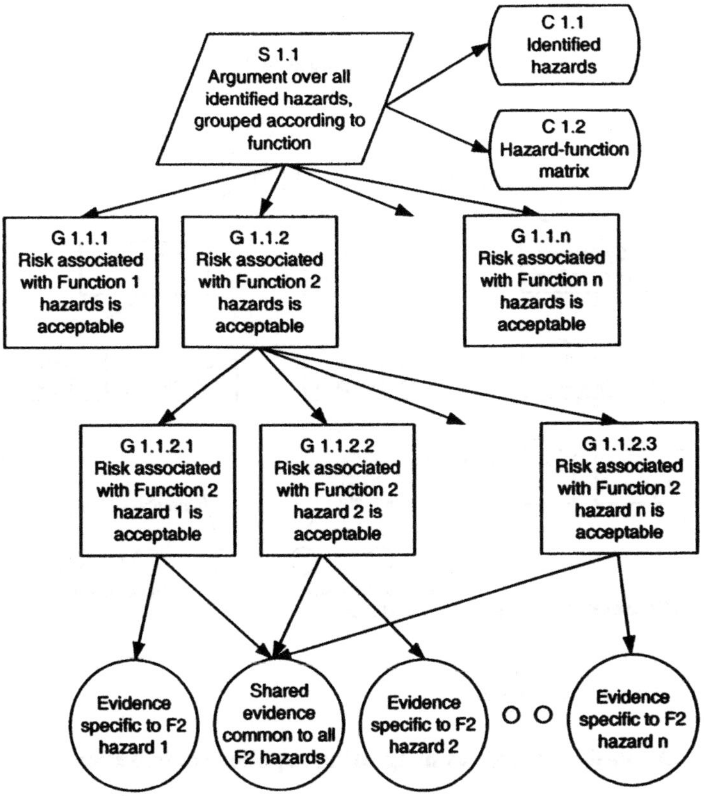

Figure 12 - Using functional breakdown to organise identified hazards

To further ease navigation of the hazard-directed argument, the GSN in the HEAT/ACT PSC document is supported by a matrix showing the functional breakdown of system hazards; this allows readers to quickly identify hazard ⇨ function and function ⇨ hazard relationships.

4.4 Incorporating supplier contributions

A final area where the structure of the HEAT/ACT safety case posed challenges was in deciding how best to incorporate evidence, or sub-arguments, provided by suppliers of subsystems and software. The first thought was to provide a single, clean "interface point", where a section of argument and evidence provided by a supplier could simply be "plugged in", as shown in Figure 13. The hope was that suppliers could very simply be shown how their work would contribute to the overall safety case, and that this would simplify the process of contracting for safety-related work.

After some experimentation, however, it became clear that a single point interface like this simply would not work. Even though a supplier might be working on a

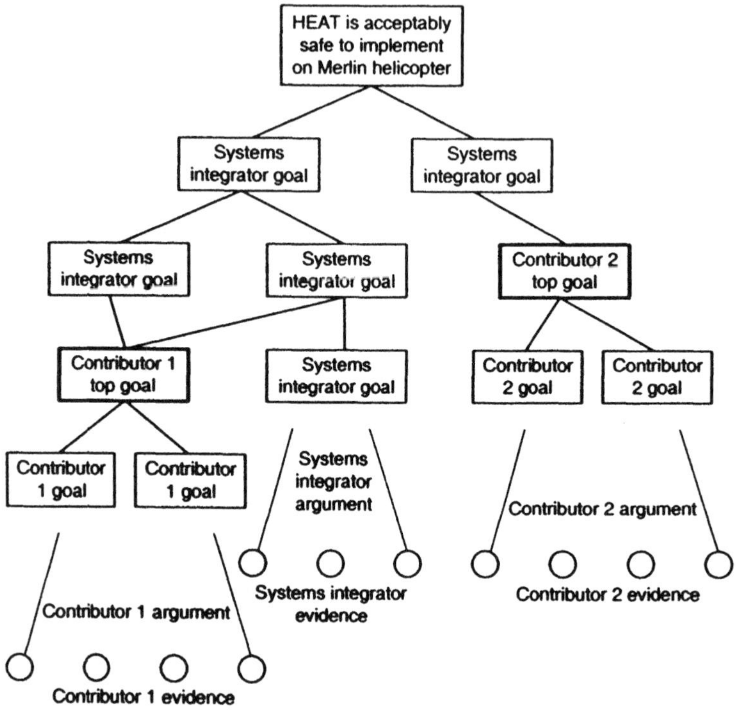

Figure 13 - Initial attempt at integrating supplier contributions to the safety case

relatively small item in terms of development, it was found that trying to pull all of their evidence contributions under a single goal led to unacceptable distortion of the argument structure. This was particularly true where there were both "product" and "process" elements to the information required from a supplier. Eventually, it was concluded that it was better to construct an "ideal" argument structure, without considering who would be responsible for sourcing each item of evidence. Provided that the PSC structure is in place sufficiently early in the project, it is possible to use this to show suppliers what they will be required to contribute, and how their contributions fit into the "big picture". Where a supplier is found responsible for supporting multiple safety case objectives, we recognised that their contributions to the overall safety case should be captured by means of an explicit safety case 'contract' between the top-level and supplier safety arguments – as discussed in (Kelly 2003).

5 Conclusions

This project has provided a convincing demonstration of the advantages of using GSN in safety case construction in an industrial setting. All of the participants in the project were impressed with how much the technique assisted in the development and, especially, the review and acceptance of the Preliminary Safety Case. Although a lot of safety work remains to be done on the HEAT/ACT project, we are confident that the argument structuring work done on the PSC will provide a solid foundation for completion of the Interim and subsequent Safety Case phases.

It was particularly impressive to discover just how powerful the use of argument patterns could be. Not only did it permit very rapid progress, but we believe that this approach has helped to create a safety case which is more thorough and robust than that which might have resulted had the whole argument been constructed from scratch. A particularly significant feature of this project has been the reuse of ideas from other safety cases, and this is an approach we would strongly recommend to anyone starting work on a new safety case. Again, the use of GSN was extremely helpful here; it was easy to review existing material, and identify argument structures that were particularly compelling, or elegantly expressed.

The project encountered a number of practical issues, which have been discussed in section 5 above. None of these were especially difficult to resolve, but it is clear that, as with writing text, it is important to think carefully about the impression that is being conveyed through the way the GSN diagrams are structured.

A practical suggestion that we would like to make outside the scope of this project is that a discussion forum – perhaps a web site – where users of GSN, and specific GSN tools, could share pattern libraries and help each other out, would be beneficial to the safety community.

Our overall conclusion is that the use of GSN and a pattern library was of huge benefit to this project, and is an approach we would recommend.

6 Acknowledgements

The authors would like to thank Westland Helicopters Ltd for the opportunity to present this work. We would also like to acknowledge the support provided by the EPSRC-funded MATISSE project (grant no. GR/R/70590/01) for some of the work presented in this paper.

7 References

Adelard, 2003, *The Adelard Safety Case Editor – ASCE*, http://www.adelard.co.uk /software/asce/index.htm

Civil Aviation Authority, 2001, *SW01 – Regulatory Objectives for Software Safety Assurance in ATS Equipment* in Part B (Generic Requirements and Guidance) of CAP670 – Air Traffic Services Safety Requirements, London, UK.

Eurocontrol, 2001, *The EUR RVSM Pre-Implementation Safety Case*, http://www. eur-rvsm.com/safety.htm#precase.

Graham, K., 2002, *Heavy Modifications: A Three Stage Safety Process for Modification of Undocumented Legacy Systems*, MSc SCSE Project Report, Department of Computer Science, University of York, UK.

Kelly, T.P. & McDermid, J.A., 1997, *Safety Case Construction and Reuse Using Patterns* in Proceedings of 16th International Conference on Computer Safety, Reliability and Security (SAFECOMP'97), Springer-Verlag.

Kelly, T.P., 1999, *Arguing Safety - A Systematic Approach to Safety Case Management*, DPhil Thesis, Green Report YCST 99/05, Department of Computer Science, University of York, UK.

Kelly, T.P., 2003, *Managing Complex Safety Cases* in Current Issues in Safety Critical Systems: Proceedings of the 11th Safety Critical Systems Symposium, Springer-Verlag.

Kelly, T.P., et al., 2003, *Hazard and Risk Management & Safety Cases,* MSc SCSE module notes, Department of Computer Science, University of York, UK.

Ministry of Defence, 1996, *Defence Standard 00-56 Issue 2: Safety Management Requirements for Defence Systems*, Glasgow, UK.

Staple, A. & Handcock, A., 2002, *The All-Electric Rotorcraft – Challenges and Opportunities*, 28th European Rotorcraft Forum, Bristol, UK.

AUTHOR INDEX

GPSR Compliance

The European Union's (EU) General Product Safety Regulation (GPSR) is a set of rules that requires consumer products to be safe and our obligations to ensure this.

If you have any concerns about our products, you can contact us on ProductSafety@springernature.com

In case Publisher is established outside the EU, the EU authorized representative is:

Springer Nature Customer Service Center GmbH
Europaplatz 3
69115 Heidelberg, Germany

Batch number: 10755053

Printed by Printforce, the Netherlands